Praise for
How the Web Won

"I doubt if anyone has had a greater impact on how people persuade, influence, and sell online than Ken. God knows how many billions his alumni have made. Nobody anywhere has had such an impact. Read this book and understand before your competitors do."

- Drayton Bird
Author of *Common Sense Direct & Digital Marketing*.
Former Worldwide Creative Director & Vice Chairman,
Ogilvy Direct

"A compulsively readable account of how the Internet became what it is today, through the eyes of one who was there and played a central part. McCarthy's writing is vivid, sometimes startling, and readable as a thriller. It's also clear-headed. This revelatory account of the web, its history, and potential, ends by reminding us: 'The Internet was created to enhance your life, not to be a substitute for living.' I suspect it will stand as a classic.

- Grevel Lindop
Poet and academic.
Professor Emeritus University of Manchester.
Author of: *The Opium Eater: A Life of Thomas De Quincey*,
A Literary Guide to the Lake District, *Charles William: The Third Inkling*, *Travels on the Dance Floor*, and multiple books of poetry.

"Fascinating, funny, and wildly entertaining. Ken McCarthy takes you through the twists and turns of the early pioneers who saw the future, and then created the biggest social and business change since the Industrial Revolution."

- Jonathan Mizel
CEO, Cyberwave Media Inc.,
Online business entrepreneur & thought leader since 1993

"An epic tale, well and conversationally told, offering deep-but-accessible insight into the evolution of our information superhighway. In the early 1990s, Ken McCarthy was part of a small group that drove the Internet's transformation into its current, advertising-funded free-for-all, winning out against the forces who'd have preferred it to stay niche and nonprofit and others who wanted to monopolize it as their private profitable niche, and he makes a strong case for why that outcome—no matter how annoying those pop-ups may get—was the best-case scenario."

- The Chronogram Magazine, Anne Pyburn Craig

"A stand-out read. McCarthy's unique perspective and experience contributing to launching the World Wide Web brings decisive insight and authority to a lesser-known subject. This slice of history hits at rarely discussed moments of ingenuity and invention, as McCarthy confronts myths about the Web's development, the evolution of its early equipment, and the funding sources that kickstarted its financial success… delivering remarkable accounts of a global phenomenon that transformed contemporary life"

- BookLife (a division of Publishers Weekly)

"There was a turning point when people first saw the Web, saw how it was all going to work, and instantly, everything changed. McCarthy was in the right rooms in the right cities at the right times with the right people to see that moment being birthed. It's like getting a peek at the early side of the Big Bang… For anyone curious how the web really came to be, *How the Web Won* provides an accurate, 360-degree view of the magic moment that will keep you engaged — and hoping for a sequel."

- Steve O'Keefe
Author of *Publicity on the Internet*,
personally tutored Jeff Bezos on selling books Online

"The story of the man who built the bridge from offline to digital marketing. No one has crossed the bridge from old-school direct response marketing to the new digital marketing world more successfully, completely, and authoritatively than Ken McCarthy. In fact, he did not just cross this bridge; Ken was one of its foremost architects and builders. Whether you are an old mail order guy like me or a GenX/Y/Z marketer, you will love and learn from this book–sage advice and deep wisdom, as lived by the author for decades in the real world of business and commerce."

- Robert W. Bly
Author, marketing consultant, veteran copywriter

"Easily one of the single greatest books about marketing I've ever read... it elevated my thinking on not just selling online but how to approach problem solving, see opportunities where nobody else does, and using history to predict what could be coming down the pike. You will notice dozens of seemingly unrelated people, places, events, and historical/corporate/technological anomalies... all mysteriously connected (think the TV show "LOST", it's almost that surreal)... where if any one of these events or conversations had not happened, precisely when they happened... or without the specific people involved... the Internet as you know it, and your ability to freely sell on it, may have never happened at all. Or if it did, it'd almost certainly be something horrifying and evil and probably prohibitively expensive everyone would despise."

- Ben Settle
Author, publisher, email copywriter

"Fascinating and hard-hitting."

- Richard Koch
Author of the million-copy bestseller *The 80/20 Principle*

"A rare look at the decisions and events that shaped the Internet into the commercial hub it is today. McCarthy breaks down complex concepts with clarity, making them applicable to modern business challenges. Plus, you'll appreciate how the biggest Internet titans today were created by ordinary people who chose to seize the opportunities before them. The book goes beyond storytelling, serving as a guide for anyone navigating today's digital economy. It reminds readers that understanding foundational principles is essential for leveraging new opportunities. *How the Web Won* is essential reading for those ready to adapt and thrive."

- Robert Skrob
CEO, Membership Service, Inc.
Author of *Retention Point* & *The Connector Effect*

"We are fortunate to have among us THE visionary pioneer who recognized what the World Wide Web really was and what it would become Before just about anybody, Ken McCarthy saw it as ad media. He was there at its birth, is respected (and reviled) as a very stubborn truth-teller about it and has done a fantastic job with this book. He tells all. I urge getting this book and reading it. From it, you can actually understand what you are using! And he has his book title correct; *IT* won."

- Dan Kennedy
Consultant & author of the *No BS* business series
(over 1 million copies sold)

"I knew Ken was a pioneer, but I didn't realize just how far ahead of everyone else he really was."

- Ryan Healy
Copywriter and entrepreneur

How the Web Won

How the

Web
Won

**The Inside Story of How a Motley
Crew of Outsiders Hijacked the
Information Superhighway and
Struck a Blow for Human Freedom**

Ken McCarthy

How the Web Won

Published by:
The System Press
PO Box 145
Tivoli, NY 12583

Version 1.2

Names: McCarthy, Kenneth, 1959- author.
Title: How the Web Won: The Inside Story of How a Motley Crew of Outsiders Hijacked the Information Superhighway and Struck a Blow for Human Freedom / Ken McCarthy.
Description: Tivoli, New York : System Press, 2024. | Includes index.

Paperback ISBN: 978-0-9773302-4-9
Hardcover ISBN: 978-0-9773302-5-6

Cover design by Taura Hanson

For permissions, please contact Books@KenMcCarthy.com

To Bettina

You are the sunshine of my life.

If I started thanking you in detail, this would be a multi-volume book.

Dedication

Until the science of geology matured to the point where it figured it out, mankind believed that salt, one of the most essential elements for human survival, was rare and available from only a limited number of sources. In the meantime, wars were fought over it and people who controlled access to it lorded it over others.

Information in the form of text, and later audio and video, were also once similarly constrained. Because of the cost of publishing books, magazines, and newspapers and broadcasting radio and television, the dissemination of knowledge was in the hands of a very small number of people. Much worthy information, painstakingly gathered and useful to humanity, never had a chance to see the light of day to the impoverishment of all.

In the early 1990s, the foundation was laid for this ancient logjam to be broken.

This book is dedicated to the many people who poured themselves into this effort long before there was even the vaguest hope of personal profit, especially the folks who labored on the project before 1995.

If, in that era, you wrote code for Internet software, configured a web server, made a web page, started a website, or even just downloaded and wrestled with your first browser for your personal computer – especially if you also inspired and taught others how to do the same – you're my hero.

"Old men forget: yet all shall be forgot,
But he'll remember with advantages
What feats he did that day.
Then shall our names
Familiar in his mouth as household words...
Be in their flowing cups freshly remember'd.
This story shall the good man teach his son;
And Crispin Crispian shall ne'er go by,
From this day to the ending of the world,
But we in it shall be remember'd;
We few, we happy few, we band of brothers"

- William Shakespeare, Henry V

It is an old remark, that all arts and sciences have a mutual dependence upon each other… Thus men, very different in genius and pursuits, become mutually subservient to each other; and a very useful kind of commerce is established by which the old arts are improved, and new ones daily invented.

- William Brownrigg, *The Art of Making Common Salt*
London, 1748

It is a splendid project and may be executed a century hence. It is a little short of madness to think of it at this time.

- Thomas Jefferson, responding to a proposal to build a 339 mile canal from Lake Erie to the Hudson River, 1809
(It was built and is now known as the Erie Canal)

When you do things right, people won't be sure you've done anything at all.

- Ken Keeler, from an episode of the animated TV show *Futurama* ("Godfellas", 2002)

Table of Contents

Part Four - Now What?

Part Five - Collected Writings and Talks

Introduction

"Man is excellently made and eagerly lives the kind of life that is being lived."

— Mikhail Zoshchenko

There are libraries full of detailed histories written about wars and battles; exploration; dynasties and historical epochs; business empires; cultural trends; and great men and women in various fields, but strangely there aren't many histories written about the media technologies that have changed and form our daily lives.

Most educated people are vaguely aware that someone named Marconi had something to do with the development of the radio and that Samuel Morse invented the telegraph, Alexander Graham Bell invented the telephone, and Thomas Edison developed the first phonograph, but they'd be hard-pressed to name the people who invented the television, motion pictures, or other key parts of our media environment like the JPEG, the MP3, or even e-mail.

They'd be especially hard-pressed to describe the path these media took from laboratory experiments with no immediately obvious use to the ubiquitous media appliances they are today.

Marconi thought the best use of his invention, the radio (the wireless radio-telegraph) would be to allow ships at sea to communicate with each other. Bell thought the best use of the telephone would be piping music to homes. Edison originally developed the phonograph as a way to relay telegraph messages automatically without the need for a human being to re-send them by hand.

One of the themes of media history is that the people who invent a technology are rarely, if ever, the people who figure out the practical uses for it that make it commercially viable, widespread, and ultimately an inseparable part of everyday life. Those insights come later and very often the people who do that all-important work are unknown. Even the fact that things like e-mail, MP3, and JPEG were *invented by someone* rarely crosses most people's minds.

This book is a memoir and is not intended to be a complete history of the Internet. (That would take a shelf full of books.)

Instead, I'm focusing on roughly three years (1993-1995) when the Internet suddenly and unexpectedly transformed itself from a government-owned network where commerce was forbidden to the commercial environment we know today where trillions of dollars in business are transacted annually.

In 1989 when the Internet was formally liberated from its federally-enforced non-profit and fully subsidized status, no one had a clue how it could pay its own way, and that cluelessness persisted unbroken until 1994 when suddenly everything came together and the path forward became "obvious."

Because this book is a memoir, contributions I made to the Internet's transformation into a commercial medium are at the center of the narrative. My reasons for doing this are threefold:

1. If I don't do it, who else is going to?

2. If people from the pioneering days don't tell their stories, future historians are likely to believe the much-skewed version of events emanating from the PR departments of companies like Microsoft and Apple, or worse Google and Facebook. Fact: In the early 1990s, Bill Gates and Steve Jobs were on the record as being contemptuous, even publicly

hostile, to the idea that the Internet could ever develop into anything significant for the general public.

3. Young people, and really anyone who is inspired to take an uncharted path, need examples of how it's done and why the walkers of uncharted paths – not the hyped-up "captains of industry" – are very often behind much of the practical progress in this world. In short, to get things done, it helps to know the stories of people who got things done. This is a story of someone who got things done.

In this book, I use the terms Internet and Web interchangeably. At the time of this writing, people are much more likely to use the term Internet than the Web. They typically say "I saw it on the Internet" as opposed to "I saw it on the Web."

Strictly speaking, the Internet is the foundation - all the machines, communication links, and software. The Web is built on top of that, like a very large and growing city with many buildings, the buildings being websites. It's the Web that made the Internet useful and usable for people.

The Web is what all the brand name services are built on. Take away the Web and there is no Facebook, no Twitter/X, no Instagram, no TikTok, no Amazon, no YouTube, no Google, and even no public-facing artificial intelligence or cryptocurrency and blockchain.

Another term I use often in the book is publishing. I define this as bringing a piece of intellectual property - an article, a book, a recording, an interview, a piece of music, a documentary, a "how to" video - to the marketplace with the intention of some kind of profit. I use it because it's more descriptive of the process than just "releasing" or "posting".

I realize that much I've what I've written so far may seem to many to be obvious to the point of absurdity, but a back-of-the-envelope estimate reveals that there are roughly four billion people living today who have no experience of the world before the Web. Understandably, to them, things like Facebook and Google just *are*, like the clouds and the trees. In many ways, this book is written for them, to give them a picture of life before the Web and let them see how the media world they live in came to be.

Does it matter? I believe yes, for a number of reasons.

First, it's generally a good principle to know what the heck is going on around you and part of that involves knowing something about how the things you deal with every day work and where they come from.

Second, while a bare list of the names of people you've never heard of is understandably not very interesting, the *story* of how these people and others came to create what they did and *how* they did it most definitely is. There were many twists and turns on the road to creating the Web.

It was not a straight line. It was not easy. The end of the story was by no means obvious at the beginning and at many critical junctures it would have been very easy for the whole thing to have gone completely off the rails - it almost did - and we would not have the Web we have today.

According to the dictionary, a memoir is "any nonfiction writing based on an author's personal memories. The assertions made in the work are thus understood to be factual." Based on this definition, I'd call this a Memoir Plus.

The "Plus" of this Memoir Plus is that while putting the book together I did some digging to make sure I spelled names correctly and got dates right. In the process, I expanded my knowledge of the details of the back stories of many of the events I was involved in, details I wasn't aware of at the time the events were unfolding. It was a fascinating process, and I learned that there are a lot of fundamental inaccuracies in the commonly told versions of Internet history.

There's also a tremendous amount of interesting and important detail that's been entirely left out, so I set out to fill in the blanks. Now that the work is done, I can say that, as far as I can tell from a serious deep dive, there is no book that covers the critical formative years of the Web's history, 1993 to 1995, in such a comprehensive way. Thus all the footnotes.

Speaking of footnotes, there are two kinds in this book:

1) Footnotes (numbered): Short notes on the page to clarify things that might not be common knowledge.

2) Endnotes (lettered): Longer notes in the back of the book—mini-chapters in their own right. Ignore them if you like, use them if needed, or dive in for more if you're enjoying the story.

The book is divided into four parts. If you want to go straight to the details of the Big Boom (1993-1995) go to Part Three.

Part One "Stirring the Pot" covers my pre-digital life in the pre-digital world where creating media was literally manual labor. I imagine young folk will look at this account in the same way we read stories about people who used to hand-churn their own butter.

Part Two "Planting Seeds" details what it was like to be in San Francisco in the early 1990s unknowingly at the cusp of a media revolution on par with the inventions of writing and printing.

Part Three "Getting the Band Together" is where I take out a magnifying glass and lay out the details of all the uncoordinated, but interlocking events that had to take place with Swiss watch precision that were necessary for the Web to have come into being.

Part Four, "Now What?", recounts stories of some of the things that happened after the Web won and it became the only online game in town.

The Appendix has three noteworthy artifacts from the early days of the Web:

1) an article I published in October of 1994, which I believe is the first article that addressed the commercial potential of email in a marketing industry journal,

2) a talk I gave on November 5, 1994, in San Francisco, spelling out the Web's likely commercial trajectory, to an audience of multimedia developers and Internet engineers, including then-23-year-old Marc Andreessen,

3) an interview I published in May of 1995 which articulated the practicalities of email publishing in useful detail.

There's also an excerpt from a 2017 *Time Magazine* article about the history of selling clicks. It's worth a look, too.

- Ken McCarthy
November 3, 2024

- Part One -
Stirring the Pot

Chapter One
Least Likely to Succeed

"Keep a green tree in your heart, and perhaps a singing bird will come."

— Chinese proverb

I am not a journalist and I'm not a scholar. And I'm most definitely not a technologist.

If you put a gun to my head and told me I had to do anything but the most basic things with my computer, I would have to start a serious contemplation of the Afterlife.

That said, I also don't know how ink is made or paper is milled and if you asked me about the details of how the physical version of this book came to be - what brand of printing press was used, the details of how it works, what one must do to maintain or repair it - I'd be equally clueless.

I can't code. I'm not even sure what it means "to code".

I didn't get a personal computer until a year after the Macintosh came along and I still miss MacWrite and MacPaint.

Most computer programs, and even many apps, leave me scratching my head at best and cursing at worst.

I surf, I send and receive e-mail, I use Zoom, and occasionally I compose an old-fashioned letter on a word processor. And, of course, I can order things and press play on YouTube and audio players. But that's it. I don't know much more and I don't want to know much more.

Despite my abysmal ignorance, somehow I've managed to get by.

In contrast, my father was one of the first generation of practical commercial computer guys.[A] He started his career in the mid-1950s and specialized in building and managing large computer systems for health insurance companies, like Blue Cross/Blue Shield New Jersey-sized systems. If I remember correctly, he once had 300 people working for him and was in the top 10% of IBM mainframe customers in the world.

It doesn't seem very dramatic now, but at one time keeping a few million individual accounts straight and mailing accurate statements was space-age stuff. I remember going to visit him at work and seeing long rows of refrigerator-sized computers with what looked like big audio tape reels going back and forth visible through glass windows. I had no idea what they were doing or how they worked. I still don't.

On a couple of occasions, I remember him coming home from work and talking excitedly about new things like random access memory. Another time, he came home thrilled with the news of keyboards that came with cathode ray tubes, something he'd recently been pitched on. What we call a computer screen now. I didn't get what the big deal was.

I typed all my high school (Class of '77) and college papers (Class of '81) on a manual typewriter. So did everyone else. The only technical thing we needed to know was how to change a typewriter ribbon. That was enough. Even though Radio Shack was selling the Tandy TRS-80 way back then, no one I knew had a personal computer.

The only classmate I knew who wrote his senior thesis on a computer, Blair Ireland, used the *mainframe* at the campus' only computer center at the Engineering Quad. He did it because he enjoyed making hopelessly complicated things work, not because it made any sense or made the process any easier.

The model back in those days was one big computer (or cluster of big computers) locked away from the public, controlled by technologists, and accessible only by "dumb" terminals that could not do a single useful thing on their own.

The only other connection I had with computers in those days was through my roommate, Stanley Jordan. Music fans may recognize his name. In addition to being a fantastic guitar player (he could already burn the paint off the walls when he arrived on campus as a freshman), having grown up in Palo Alto, he had a fascination with computers, electronics, and computer music.

In those days, not only were computers rare and distant so was software. Princeton did not own any music software, but Columbia University did, so Stanley would program his music in Princeton, send it via a pipe to Columbia where it was processed, and receive the output via a device that transferred it to an audio cassette. The process took weeks, (maybe even months) to generate about 30 seconds of music and you couldn't be sure what it would sound like until you heard the output.

The "pipe" Stanley used to access the Columbia University music software from central New Jersey?

I learned many years later it was the ARPANET which later became the Internet on January 1, 1983. So theoretically I was aware of the Internet back in the 1970s, but I was entirely clueless about it.

My next encounter with computers would not take place until five years later, but first I did a deep and practical dive into the realities of *human* reading, memory, and cognition.

A year after graduating from college, I spent a year on the road with a company that provided speed reading and study skill classes to private colleges. We'd land on a campus, teach for four weeks, pick up stakes, and then go to the next place. I got to see parts of the country a Yankee like me might otherwise have never visited like Due West, South Carolina (Erskine College), Salem, Virginia (Roanoke College), and Ashland, Virginia (Randolph-Macon College).

It was a lonely job. I'd land in places where I didn't know a soul and I had a lot of time on my hands. So I read – a lot. The first thing I'd do when I got to a new campus was go to the library and take out every book on reading, memory, study skills, cognition, and rhetoric and read through them all in an attempt to make my classes more interesting and effective.

Every college library seemed to have its own unique collection of books on these subjects, with each book coming at it from a slightly different angle often with unique sets of facts. In the process I learned an important lesson: If you want to learn a subject, don't just read a book about it, read *shelves* of books about it. If you do that, there are very few subjects you can't become a legitimate armchair expert about in six months or less.

This was reinforced for me later by Eugene Schwartz, the scholar/ad copywriter/publishing entrepreneur/art collector. Before he sat down to write an ad about anything, he not only knew everything about the product but also everything about *all* the products in the market – past and present – and as much "peripheral" information as he could find.

One of my favorite sayings is "The mine is bigger than the gem." You never know what you'll find if you dig through tons, not pounds, of earth. I can't think of a single time in my life when the process was ever a waste of time.

It's not unheard of for people to read entire libraries. For example, Harry Truman, while he had some education after high school, never got a college degree, but managed to read every book in the Independence, Missouri library. His conclusion: "Maybe I was a damn fool, but it served me well when my terrible trial came." In his case, the terrible trial was suddenly finding himself President of the United States in the middle of World War II.

If I remember correctly, the course we taught had twenty days of material – five days a week, one hour a day, for four weeks. I'd teach the same class four times a day which to me was a great thing because I could continuously refine my presentation and make it clearer and more impactful. After six months of this process, I had developed a very strong class, one which bore almost no relation to the one I had been trained to teach.

The big thing I learned is that virtually everyone, no matter how intelligent or well-educated, has some trouble with reading, sometimes physical, sometimes cognitive, sometimes both, and that school systems, either through ignorance or sloth or a combination of both, don't teach students how to identify these problems and overcome them. They can all be overcome.

When I taught this subject later in New York City, I had a student who had had twelve unsuccessful eye operations and was only able to read about 80 words per minute which is beyond slow. Despite these physical limitations, he loved books (he even made his own) and was whip-smart. Within a month, he was reading at a normal pace of about 350 words per minute. I just helped him retrain the way he used his eyes.

The main cognitive problem many people have when reading for information is that they mistakenly believe reading starts at Word One and carefully read each word to the last word. This may make sense if you're an attorney reading a contract for a client, but it's a terrible strategy for getting the meat from a book. Reading non-fiction (I know nothing about reading fiction) is all about information extraction and organization.

Books, if they're written halfway decently, are structured packages of information. Your goal as a reader is to methodically go through the "Big Store" of information that a book represents, identify the different departments, and find the interesting and useful "products" in each department. If you methodically do this, you will automatically read faster *and* remember more. These two positive outcomes are *not* contradictory.

Unlike a computer, which eventually maxes out on storage, the more structured information you put into a human mind, the bigger the capacity for the mind to absorb new information becomes. Long ago, I wrote a book about all this and I recently discovered the manuscript in my attic. I will be putting it between the covers of a book at some point in the not-too-distant future.

While I was at my last teaching assignment, Moravian College in Bethlehem, Pennsylvania, I decided to create a one-day seminar version of my own for the students at Lehigh University, another college in town. I rented a space, put up flyers all over campus, took phone calls (no Internet then), and when the day of the seminar came, collected people's cash and checks (no credit cards for college students then either).

The seminar was a success. I just made one mistake. When I got to the classroom, I realized that I'd left my teaching notes at home which was 90 minutes away. So I just walked the students through the notebooks I'd prepared for them and did it all from memory. It worked and I gained new faith in my ability to improvise and keep a roomful of people entertained for a day.

Previous to my year on the road teaching, I was not a public speaker. My limited experience speaking to groups was something out of a cartoon. My knees knocked against each other as I stood there in terror. I'd heard the expression of what the knees did in a panic situation, but I didn't know it was a real thing. I can tell you from first-hand experience, it's a real thing. Another real thing: cotton mouth. How can a mouth go so thoroughly dry so quickly?

Anyway, I got over it. It's a very good thing to get over, as painful as the process may be for a bit. Now, speaking to groups of any size is like riding a bicycle to me, fun and good exercise.

It occurred to me that the material I'd gathered would be very helpful to the students at the college I went to. It had no such program.

Princeton gets a lot of kids from prep schools, short for college preparatory, where they are specifically trained to be prepared for college-level work. They arrive fully at home knowing how to take useful notes from complex lectures, tackle difficult reading, engage in give-and-take discussions, do research, and write multi-page essays.

But Princeton also gets some kids who are not up to speed on these things. When I was a student, I clearly remember one young woman with an impressive but misguided initiative who learned shorthand so she could take down every word of every lecture, transcribe them, and then read and study the transcriptions. No one had ever taught her how to take notes.

I contacted Princeton, and told them about my experiences teaching reading and study skills to college students, and offered to help. And lo and behold, I was invited to give a workshop during Freshman Week and I was told that if it went well it might lead to developing some kind of formal program. Based on the incoming students' reaction to the workshop which was very positive, I assumed the people who were getting paid to help new students adjust to college life would want to continue the momentum and see that other students were helped. For some of these freshmen, starting at Princeton with the preparation they had *not* received was like jumping out of an airplane without a parachute.

The administration certainly made noises like they were interested in the idea, and in my naïveté I took those noises to mean something, but as I learned as the months drew on, and my savings diminished, they were just paying lip service to the idea. They had no intention of developing a skills program for students with me or anyone else.

As best as I could figure, the idea of teaching Princeton students something so common as how to study was somehow "beneath" the school and wasn't congruent with its image of itself. (Better to let the kids flounder needlessly I guess.) So instead, I started teaching local high school students and their moms were more than happy to pay me to help their kids learn how to study.

Then I saw an ad for a job paying $15 an hour to do something called technical writing. I had no idea what technical writing was, but I knew I could write and at that time $15 an hour which worked out to $30,000 a year which seemed like a fortune. By way of comparison, my classmates were making on average $18,000 their first year out.

With high hopes, I applied for this dreamlike position, and much to my surprise, I got the job. For a few weeks, the work was about 15 minutes from home. Then I got the word that I was needed in Fort Lee, New Jersey.

Fort Lee, New Jersey is 60 highly urban miles away from Princeton. Not the end of the world, but I had a car that probably only had about 1,000 miles left on it if I was lucky. So I drove to the train at Princeton Junction, took the train to New York City, took the A Train to the uptown Port Authority bus terminal near the George Washington Bridge, took the bus over the bridge to Fort Lee, and then walked to my office. It took well over 2 hours each way. Up at 5 AM. Home at 8 PM. Wash, rinse, and repeat. But, hey $15 an hour and I had a lot of time to read on the various trains, when I was awake.

The job was "interesting."

The year was 1984. Personal computers were not in every home and were not in every office. Microsoft hadn't gone public yet. That wouldn't happen for two years. Microsoft Word was floundering. Sometime around then I stumbled on Bill Gates personally pitching Word in a newly-minted Egghead Software store in Manhattan. I was going to say hello, but he looked so sickly I was afraid I might catch whatever it was he had and my overriding thought about him was, "That poor guy. He's not going to make it." 'Make it' as in live very long.

In 1984, most industrial strength word processing was done on Wang, a line of computers named after the company's co-founder, An Wang. It was a single-use dedicated machine. It did word processing and that was it. IBM PCs and PC clones were already seriously nipping at Wang's heels in '84, but the machine still had a large installed base and came with a service contract. Corporate managers worship at the altar of inertia, and computer repairmen in those days were few and far between, so Wangs persisted.

Being a Wang operator was a good gig, especially if you lived in New York City. Corporate law firms and financial institutions produce mountains of high-level documents every day and there was a serious shortage of Wang operators. Thus, for a brief shining moment, people who could type and knew the system not only made $25 or more per hour during their post-5 PM shifts (a fantastic hourly wage in 1984 the functional equivalent of $75 per hour today), but they also got dinner delivered to their desks (lobster and steak were not unheard of) and car service home at the end of the night. It was a special boon to actors, musicians, and other creative people who, if they landed a creative gig, could afford to take it knowing they'd always be welcome back later – and at $25 per hour – until word processing became democratized and they weren't.

I was not a Wang operator, but I was intrigued. After all, $25 an hour is better than $15 an hour, but it looked too complicated to me. Instead, I was a proofreader and copy editor. I would do things by hand using a pencil, as everyone else did, give it to the Wang ladies, and they'd do their Wang magic.

But there was just one problem. When I checked their work, they would make the changes asked for (most of the time), but somehow they managed to create brand-new and quite random errors. (These were *not* the $25 per hour operators. They were former typists who were struggling with the new technology.) We were working under incredibly tight deadlines and the responsibility – and the pressure – of getting these documents done and ready for the typesetter on time fell on me.

The owner of the tech writing company, the guy who went out and got the business, wisely had his office on another floor and never set foot in the production office. My theoretical $30,000 a year was grossing him $30,000 a year too because he marked up our invoices 100%.

There were twenty of us on this project and this was not the only crew he had working. Do the arithmetic. Of course, he had rent, utilities, insurance, the Wang workstation rental fees, and probably kickbacks to the corporate people who threw him business, but clearly he knew how to shake the money tree.

The guy who was supposed to manage our project and troubleshoot it rarely made an appearance in our office and when he did he was twitchy and fast-talking. His purported job was to supervise the company's production offices in various parts of Northern New Jersey. My guess is he spent a good part of his day powdering his nose (which for people who don't know the slang means doing cocaine, which was a popular activity at the time for people who had more money than brains).

What exactly was the job I signed up to be part of?

The client, AT&T, had decided they wanted to go into the personal computer business. They also wanted to have proprietary software. They had none and didn't have the capacity to produce software for PCs so they acquired the software from another company.

In those days, all software came with printed manuals. The only manuals for the software AT&T acquired were stamped with the original company's logos and copyrights and had to be redone from scratch. This was in the age before reliable optical character recognition, so effectively, they had to be retyped from scratch by hand and reformatted with all the appropriate AT&T logos and copyright notices.

There were three problems with this venture:

1) The deep thinkers at AT&T were very late to realize that they needed manuals to accompany their PC clones. Thus the Glorious Launch they envisioned of putting an AT&T logo on PC clones and charging a premium price for them was stuck on the launch pad.

2) Amazingly, we, the people who had to produce the new manuals, on short notice and under high pressure, were not allowed, for "security" reasons, to have access to either the machines or the software to use as a reference to confirm that the instructions in our new manuals made any sense.

3) The manuals from the original software company that we were given to work from and adapt for AT&T made no sense. To say they were utter crap would be charitable.

Our mission:

> Using garbage manuals as our reference and having no access to the software or the machines it ran on, to somehow intuit how the software *should* work, make the fictional manuals look good, and get them out the door on time so that the deep thinkers in the C suite could have their epic launch.

If you've ever wondered why the user manuals for products you've spent your hard-earned money on are often useless to the point of worthlessness, this story pretty much sums up how manuals get made. At the last minute tech writers receive incomplete, and often inaccurate, information, with no way for them, short of corporate espionage or telepathy, to get access to the information they need to do the job right.

AT&T had some dealings in the computer hardware and software market, but my understanding is that this particular personal computer venture sunk without a trace.

After about three months of this "work," I started to mentally come apart at the seams. This was no job for a perfectionist.

The owner of the tech writing company with his office downstairs was never troubled by any of this. He had his eyes on the prize and for him, that meant incoming checks from his corporate clients. If his assistant, the guy with the well-powdered nose, had any concerns about the situation, he kept them well hidden. In my exit interview, when I pointed out the structural impossibility of the situation, the twitchy fast-talking guy said, "Look. We paid you $15 per hour." And so they did.

None of this left me with a good feeling about the computer industry.

Chapter Two
Bright Lights, Big City

"They are ill discovers that think there is no land when they can see nothing but sea."

— Francis Bacon

Before I left for my adventure traveling the South and teaching study skills to college students, I managed to snag an apartment in New York City.

After World War II and right through the 1970s, apartments were so readily available and so cheap, especially if you lived "downtown," that artists moved to New York City because of the low cost of living. Hard to believe but true.

Sometime between the time I graduated from high school in 1977 and graduated college in 1981, someone flipped the switch and it all changed overnight. Apartments, at any price, became very hard to come by and were very expensive.

Getting a good apartment with a low rent was always tricky and usually involved "key money" – paying off the super, if the building had one. (Superintendents live in the buildings and maintain them for the owners. They wield enormous practical power.) By 1980, getting even a bad apartment was considered a major triumph and if you got one at a reasonable rent, it was the equivalent of winning the lottery.

Well, I won the lottery.

A friend and I were driving on Canal Street in New York City, and I saw another friend of mine, the photographer Beth Cummins walking down the street, very pregnant. I hadn't seen her in a long time and jumped out of the car to say "Hi" and give her my well wishes. When she said that she and her husband, the percussionist Jerome Cooper, were moving to a bigger place, I asked her, "What's going to happen to your apartment?"

She hadn't given it any thought and most importantly hadn't mentioned the move to anyone who needed an apartment. She gave me the name and number of the owner.

The owner was part of a trio of brothers. One dealt in fuel oil, one was a real estate lawyer, and the third, a plumber, was the owner of the apartment I was interested in. They were minting money and starting in the 1960s bought every apartment and brownstone that they could get their hands on when they were cheap. By the time I came along, they had hundreds of units on the Upper West Side, which conveniently for them had evolved from a war zone into an upscale and very in demand neighborhood.

All of their buildings were in good shape – but one. Some even had doormen and one building was so upscale that one of their tenants was a star player for the New York Rangers.

However, they had one dog left in their portfolio that they never bothered to renovate, a four-story building that when built had a grand apartment on each floor. During the bad old days, before they bought it, it had been roughly cut up into 16 tiny studio apartments. The tenants were interesting. There were a lot of young people like me, but there were also a lot of people for whom this place was their last stop. One guy occasionally would get drunk and fire off his .45 automatic. He didn't do it very often and luckily it was old-school construction so the walls and the ceilings were pretty thick.

My apartment was on the top floor and was about 10' x 14' feet with enough room for a queen size captain's bed (a bed with drawers underneath it), a table less than half the size of a card table, one folding chair, a small closet, a kitchen about the size of what you'd expect on a very small sailboat, and, most important to me, a floor to ceiling bookcase that separated the bedroom from the kitchen.

I loved it. I especially loved the rent. If I remember correctly, it started at $248 a month which was frankly a miracle.

The bathroom was all mine, but it was in the hallway and had no sink. I was on the top floor, which meant in the summer the roof heat would turn my room into an oven. For the first three years I was there, air conditioning seemed like an extravagance – until the night I woke up and found myself on all fours in the hallway, panting. It had become so dangerously hot in my apartment that my body said "Get out of here!" and dead asleep, I somehow got out of bed, unlocked the door, and made my way to the hallway.

Other highlights: My next-door neighbor caught a rat in his room the size of a small dachshund.

On the positive side, it was in the middle of the block, and believe it or not, being in the middle of a block in New York City can be as quiet as living in the country, and mostly it was. I had two big windows that let in lots of light. I had room for my books and most important I didn't have to sweat the rent, but shortly after getting this gem and cutting my commute to Fort Lee from two and a half hours to 50 minutes, I was fired from my tech writing job.

OK, now what?

I knew how to teach speed reading and study skills and here I was in a city that never sleeps, has a lot of readers, and a lot of places to post posters. So I hand-made a poster with my phone number on it and started posting it in search of students.

Very quickly, I realized I needed a better poster so I looked into getting it typeset which was the only way to get posters made in those days if you didn't have a personal computer and a laser printer, which I did not. So I worked up a new poster, agreed to the $100 typesetting fee for one page, and left it off to be done. When I picked it up, I realized it had a typo. (Did I mention how much $100 was in those days?) They were disinclined to give me a do-over.

After this experience I made the acquaintance of a laser printer and the MacIntosh computer which happened to be in the computer lab of the campus of Columbia University. Luckily things were looser then and I looked like a Columbia student, and nobody was checking too carefully if I was or wasn't, so I commenced introducing myself to the marvels of desktop publishing.

It's hard to imagine today what a breakthrough this now-ubiquitous technology was. By way of comparison, if in the era of pre-desktop publishing, you had the audacity to self-publish something, you had to type it out on a typewriter.

Homemade posters before desktop publishing were drawn by hand or if you were super-sophisticated, you'd use stencils. (Even professional architects made their blueprints this way.) The super-duper sophisticated had something called Letraset which were sheets of all the letters of the alphabet, upper and lower case, and in various fonts and sizes, that let you hand transfer a letter from the sheet to the page you were working on by careful rubbing. Even the Pentagon and Exxon didn't have anything better, though they did have Selectric IBM typewriters which had removable type balls that gave you access to different typefaces including the letters of foreign languages.

With my access to a Mac and laser printer, I was free from typesetter tyranny and was able to continue to tweak my posters to make them better.

I had a very simple business model. I would put up posters on both sides of Broadway from Lincoln Center to 96th Street. Then I'd jump up to the Columbia University area and put my posters up on Broadway there and on the campus. I'd order a couple of thousand posters per month and every couple of days I go out with a backpack and make sure every bus stand, light pole, and bulletin board had a fresh one.

This simplistic marketing program was enough for me to fill a course every month at $195 per student. It was enough to easily pay my rent, eat, and buy books.

If I had been ambitious, I would've set up a separate poster campaign on the Upper East Side and one in the Village downtown and had teaching locations in all three places. I also would've created continuation courses, like "Advanced Memory Skills" and I would've followed my next-door neighbor Guy Polhemus' excellent advice and created a home study course. I did none of these things. I had zero interest in working anything close to full time and preferred to spend my time wandering the city, listening to music, and most of all browsing bookstores and reading books. My goal was to be a free man in Manhattan and I accomplished that.

However, I did take Guy's advice on one point. I perused the 1,000-page+ *Dartnell Direct Mail and Mail Order Handbook* which he kindly loaned me and it got me thinking about the importance of things like headlines and response devices. I also came across a book by Robert Bly that introduced me to the idea that there was something called "copywriting". Another important seed was planted then, but didn't bear fruit until many years later. I learned that there was a writer named Gary Bencivenga who was paid $25,000 dollars (in

1980s dollars, for the sales letters he wrote. That information was so mind-boggling I couldn't process it at the time, but it made an impression on me.[1]

When my flyer reached its pinnacle of sophistication it was one big two-word headline ("Speed Reading"), with three bullet points, my phone number, and six pre-cut (by hand) tear-off squares at the bottom with the words Speed Reading and my phone number on them.

When people called, I'd talk with them about their situation, what speed reading was and wasn't, and how it could help them. It was always a conversation and never a sales pitch, though, of course, I was selling all the way in the sense that I knew I had a very good method which I taught very well. I'd tell them about the next upcoming course and offer to send them information, by mail of course because there was no other way. Most said yes and got added to my handmade database. I'd mail them immediately and then every month when I had a new class, I'd mail them again. When all the dust settled, the outputs (the money I banked) were significantly bigger than my inputs (the time and money I spent) and enough to live on, most of the time. That was the whole business.

As I learned over the decades that followed, from the significant money I spent on ads and the countless millions of advertising spends I've advised on directly and indirectly, marketing doesn't get much more complicated than that. This experience and the insights that came from running that simple little business were to have a big impact on my life and the commercialization of the Internet.

---

1   Little did I know that the $25,000 dollars Bencivenga was paid was merely an advance against royalties. If his letter was successful, and many of them were spec acularly successful, he earned another 5 cents per letter mailed. This may not sound like much but one of his letters had such broad appeal and was so popular that it was mailed 100 million times. Much to my delight, over twenty years later I had the opportunity to meet both Gary Bencivenga and Robert Bly.

Most of my friends from college had the good sense when they graduated to get real jobs. And most of them ended up working for Wall Street banks which was the place to be in the 1980s. Two did so well that they both retired around the same age, forty, on a mountain of cash. One ended up with a 5-acre estate with two fine houses in Chappaqua, New York and he wasn't stretching himself to buy it.

One day, a friend of a friend mentioned over dinner that he needed to find a tech writer for a project he was working on at Bankers Trust. At that time, and it still may be true, Bankers Trust was the biggest private trader of foreign exchange in the United States. They were developing computer workstations and trading systems for their traders with the hope of someday eliminating traders and having computers do all their trading for them. I got the job.

Being able to see live prices and make trades on your home computer is no big deal now, but in those days, a trader workstation with six screens, a Telerate connection, a connection to live prices, and a super-jacked telephone system, ran about $50,000 in 1980s dollars. Bloomberg was around then, but was in a fledgling state and we didn't use their terminals.

This was the era of Andy Krieger, a name old-time foreign exchange traders will recognize, assuming there are any old-time foreign exchange traders left. The legend was that he could dial his trading partners – private and central banks all over the world – with either hand, without looking, faster than speed dial.

Why did such a skill matter? The foreign exchange market, being human and non-computerized then, would occasionally offer up situations where you could make a million dollars or more with no more difficulty or risk than picking up a twenty-dollar bill at your feet. The catch was you had to continuously keep your eye on numerous balls in the air and get in and out before your fellow sharks saw what you saw. The potential pay off was huge. One year Krieger made the bank $300 million dollars with little more than a desk, a phone, and some computer screens.

Working with foreign exchange traders taught me an important lesson about the need for speed in business: Windows of opportunity open and close fast. That understanding, combined with my "discovery" of the rudiments of direct marketing, has been worth millions to me and a whole lot more to my clients.

I find that the need for speed is the hardest thing for most people to grasp. Folks love learning techniques, but they don't seem to love what's required to implement them at a pace that makes a difference. Me, I love speed. I ran track as a kid (100-yard dash) and played basketball. I loved fast breaks, a technique in basketball in which you get the ball down court after a rebound as fast as humanly possible while the other team is still standing around adjusting their jockstraps.

Working at Bankers Trust as a contractor introduced me to the lucrative world of doing tech writing for New York investment banks. Banks, unlike tech companies, actually are required by law to accurately document their computer systems. Screwing up this function can easily result in millions of dollars in liabilities. So the banks paid up and just as it was great to be a Wang operator in those days because the demand was high and supply was low, it was even better to be a tech writer with investment bank experience.

I started at $40 an hour. Remember my rent was less than $300 a month so I could make my entire rent on the first day of the month. Earning the rest of what I needed to live on – food, dry cleaning, and subway tokens – was a breeze.

When the project at Bankers Trust ended, I knocked on the doors of contractors who specialized in financial tech writing, and they were delighted to see me. It was a new experience in my young life. I was immediately placed at First Boston, which in those days, before it shot itself in the foot, was one of the most prestigious investment banks in the world.

One of the ways it shot itself in the foot was instructive. They decided to invest in creating a brand new computer system for the bank from scratch. When you realize all the activity that goes on in an investment bank (or did then, all they seem to do now is fleece the public), they had to manage a lot of data, literally the current price of everything on earth from tea in China to eggs in Afghanistan, all the stocks on all the world's exchanges, all the bonds in all the jurisdictions, and every other financial instrument you can imagine.

I got the job of working on the Product Master which was the database of every single financial instrument the bank dealt in, which was a lot. To be the tech writer on this job, I had to become familiar with the details of all these instruments, which gave me the functional equivalent of an MBA in finance.

In any event, this ambitious program of building a brand new bank-wide system from scratch was fantastically expensive. (It's more expensive to build a new house than patch the roof.) The idea was that once developed, other banks and finance-heavy institutions like insurance companies would come to buy the system. They had one client, Mass Mutual, and I don't think they ever found another. A year after I left, one of the traders made a fantastic financial misstep that hurt the bank so badly it had to sell itself to Credit Suisse and all that development money went down the drain in an eye blink.

Waylon Jennings and Willie Nelson wrote a song called "Mammas Don't Let Your Babies Grow up to be Cowboys" to which I'd add the line, "Or software entrepreneurs either." Developing worthwhile software, marketing it successfully, and enduring the endless whipsaws of the economy and of tech itself is challenging. It's not a coincidence that the people who succeed tend to be maniacal. I'm not sure if there's any other way to survive in that business.

Something else happened while I was at First Boston. The 1987 crash. Our offices were in Midtown just off Fifth Avenue. I was meeting a friend, Russ Weiss, a third-generation cotton trader, for lunch. When we connected on the street I mentioned that the Dow was down 500. He looked at me and said: "Are you sure about that? That can't possibly be right." It was and the mood at First Boston changed dramatically after that. I remember one woman had been telling me that she and her husband were planning to buy an island in Florida and they weren't necessarily high-level people at the bank, but in those days everyone was getting paid. After the crash, her plans and a lot of other plans like them were put on hold.[2]

First Boston, and all the investment banks, stopped developing "new and improved" computer systems at the rate they had been funding them before. The "drunken sailor on shore leave" period was over and tech writing work became harder to find. Windows open, windows close. Fortunately, I'd saved nearly every penny I'd earned and had a large cushion of savings to ride me out for at least six months, maybe a year if I was super-frugal, an easy thing to do when your rent is less than $300 a month.

---

2   I don't know where to put this next story, so I'm going to squeeze it in here. When I started as a contractor at First Boston in order to orient myself to the company I went down to the video viewing room (there was only one) and watched a VHS of the CEO giving new MBA hires the lay of the land. He told the story of being on an elevator and overhearing a junior banker bragging about the details of a deal he had just closed. He stated his extreme disapproval of conversations like this in a public space: "Don't do that. Nobody needs to know how we're making our money." That was instructive.

Nature abhors a vacuum and as luck would have it one of Guy Polhemus' friends, Bill Markle, was in a bit of a pickle. He was a veteran filmmaker who got his start by teaching himself how to repair film projectors and realized that projectors used the same mechanism as film cameras. He leveraged that into becoming a camera operator, then an in-demand cameraman, and finally a director of commercials and industrial films.

All this happened during one of those golden periods when demand for skilled people far outstripped the number of skilled people available. In addition to the aesthetic part of making films, making films had a lot of demanding, no-bullshit, manual skill components to it. For example, film is edited by hand. There was no other way to do it. Film was expensive and developing it was even more expensive. Cameras were expensive and heavy. You needed to be more than a little sturdy to lug one around. There was no grabbing a SmartPhone, shooting a lot of stuff, and figuring it out as you went along at zero cost.

Did I mention video didn't exist when Bill started his career and digital video didn't become a thing until the late 1980s when he already had twenty-five years under his belt?

In 1988, it was fantastically expensive to get into the digital video business. I remember one tenant in the Film Center Building, where Bill had his office, sunk $250,000 (in 1987 dollars) into building a digital video suite, which for some odd reason we called "online editing" in those days. There was nothing "online" about it. Studio time was $100 per hour and that's what it cost me to add two screens of text to the end of a promo video that a fashion photographer had hired Bill to edit.

You may not have heard of this photographer, Marco Glaviano, but you probably know the name of the then-obscure model just signed to Elite that he used for his video: Cindy Crawford. Sometimes you get lucky. If you go to

Glaviano's website today, guess what pictures you'll see on his home page thirty-seven years later?

By the way, in the 1980's the Film Center was the ghetto of NYC's film industry on the then-very rough Ninth Avenue. Crack, prostitutes, and crack-addled prostitutes were a significant neighborhood industry. It was in a bad part of town at a time when New York City was not the safest place in the world. But it had the virtue of low rent and a great little diner on the ground floor.

Leon Gast was one of the Film Center's tenants in those days. If you're up on the winners of the Academy Award for Best Documentary in the last thirty years, you know his name. If not, you may recognize his film's title *When We Were Kings*. If you don't recognize the title, you will surely recognize the film's subject, Muhammad Ali.

In 1972, Gast made a documentary for Jerry Masucci's Fania Records, the iconic salsa label, called *Our Latin Thing* (*Nuestra Cosa Latina*), documenting Fania's stars performing at Cheetah, then a major Latin music venue of international import. This led two years later to Leon getting hired to document the Fania All-Stars' performance at the "Rumble in the Jungle", the fight between George Foreman and Muhammad Ali in what is now the Democratic Republic of the Congo.

Gast wisely shot a *lot* of film in addition to the performance he was paid to capture, and that included much footage with Ali before and after the fight, and he ended up with the rights to all the film he shot. That was 1974.

When I first encountered Leon in 1988, he was the laughingstock of the Film Center Building which was a little bit like being the lowest man on the least prestigious totem pole on the wrong side of the tracks in the worst town in the state. Why? Because he was forever talking about this film that he was working on that never seemed to get done and,

14 years later when I met him, it still looked like it was never going to get done.

Well, it did get done in 1996, twenty-two years after Gast shot the original footage, and he won the Academy Award for it that same year. Wins are everything in the film business and Leon spent the rest of his days seeing all his future projects readily financed and enjoyed a never-ending stream of praise and paid festival appearances including winning the lifetime achievement award at the Golden Door International Film Festival in 2012.

Three things to learn from this: 1) don't be so fast to throw in your cards, when you're holding an objectively strong hand (i.e. the unique Ali footage he had), 2) if you have a strong hand, put in the ear plugs to the comments of others and march on, and 3) ideally, you want to move things along a wee bit faster, but late is always better than never.

Who was one of the people who made *When We Were Kings* possible? Bill Markle.

To go back to the pickle Bill found himself in, when the mid-1980s rolled around video was on the scene, and film schools were cranking out kids who could point a camera by the container shipload. Cameras were lighter and editing video was not a skilled trade, the way cutting film – literally *cutting* film – was. Anyone with decent judgment who could push buttons could become an editor. Did I mention that thousands of kids were being trained each year to use these things? The previous supply and demand situation, which had been in the favor of film craftsmen, collapsed.

Bill took the lemons of the video and impending-but-not-there-yet digital video revolution and made lemonade. His father had been in the radio business (if I remember correctly) and his brother Gil, in addition to being a successful entrepreneur, owned Long View Farm Studios, which recorded a whole lot of people and provided rehearsal space to a whole lot more including Stevie Wonder and the Rolling Stones. So Bill knew audio. Also having made hundreds of industrial films and commercials, he knew that the coming digital revolution (which was obvious to very few in the 1980s) was going to have a huge impact on audio post-production, the stuff that's done after the film is shot to make sure the audio "works."

It turns out that audio is even more important than the pictures on the screen. Viewers will forgive less-than-perfect photography. They will even call it "art". However, when it comes to sound, the slightest audio imperfection jars viewers out of the all-important suspension of belief. They're suddenly aware they're watching a movie instead of being immersed in the experience, and that leads them to think the movie sucks. If you've ever seen one of those badly synched Godzilla movies, you know what I'm talking about. There are countless other ways, many of them subtle, to get audio wrong.

Bill tried to explain to me how complicated it was to produce soundtracks (dialog, music, and sound effects) for films in the analog days, which we were firmly still in 1988. I couldn't follow it. Editing film, I could grasp, at least theoretically, and I'd seen him edit film when a job called for it (he was fast), but the audio part which he described to me sounded like the least plausible technology ever invented. More of a bad joke than useful. The process of adding sound to film was called "optical sound." There's currently a video on YouTube called *Classical Optical Sound Recording* and you can see the process for yourself if you're so inclined.

Bill had an idea for how to do audio posting better. An edit shop in New York City had created the city's first and at that time only digital audio post-production studio. The company, which was a large film and video post-production facility, had sunk millions of dollars into it and it was very expensive to use, something only for high-budget feature films and ad agency productions for top-drawer clients.

Bill had the idea to hook up my Mac 128K, which he upgraded to a honking large 512K, to a Sony 3/4 inch U-matic video editor (think VHS with cassettes the size of hardcover books), to an Otari reel-to-reel tape machine, to a mixing board. How all this worked and how Bill knew it would work, I have no idea. It seemed like the most unlikely thing on earth and yet with this low-budget rig, he was able to provide audio production as good as a cutting-edge, $1 million-dollar-plus audio posting studio.

So what was the problem? The problem was that Bill had invested a lot of time in figuring all this out, but hadn't developed any marketing for it. It wasn't easy or obvious to explain to filmmakers of the time, who were 100% analog, what digital audio posting was. After all, they had something that already worked and knew how to budget for. Why should they stop what they were doing, take a gamble, and learn a whole new way of working? Filmmakers were not beating a path to his door.

I didn't fully understand the significance of what he had done either, and I was hard-pressed to explain the technology to anyone else, but I knew that if we got a large stream of inquiries from curious people then eventually lightning would strike and we would start booking some business. But how to reach them?

In those pre-digital days, there was something called the "Yellow Book", also known as the NYPG or the New York Production Guide. It still exists, but it's digital of course. It's the Yellow Pages for the film industry. If you're making a film and you need something, it'll be in the Yellow Book: sound stages, equipment rental, prop and costume warehouses, editors and editing suites, cameramen, lighting directors, producers of all manner of motion picture content, and on and on.

Undaunted by my ignorance and bolstered by my experience of using the mail to sell my speed reading classes, I got a copy of the yellow book and typed all the addresses into a program which was called something like My Mailing List Manager. It was a beautifully simple piece of software with fields for name and address and a print button and function calibrated for mailing labels with sprocket holes so that you could print them out on a dot matrix printer.

I put a headline on a postcard with reasons why producers would want to know more about this technology, and offered them a free report on the wonders of digital audio post-production. The well-heeled big boys ignored us of course, but there were enough producers with low budgets and big dreams that we booked $14,000 in business within 30 days of the postcard being mailed.

One of the groups that walked through the door was a Mexican production company trying to finish a feature film. The Mexican peso had just completely collapsed in the foreign exchange market and I counseled Bill not to put too much faith in them or extend them too much credit. I don't know whether he listened or not because I moved out to California before the deal was closed, but the film they were working on was *Como Agua Para Chocolate*, also known as *Like Water for Chocolate*, which turned a $2 million budget into $21.6 million in the box office, making it, at the time, one of the highest grossing foreign films of all time. If you ever see the film, every name in the credits is Hispanic, until it comes to the "Audio Syncrado" credit which goes to Bill Markle. You can't finish a movie without locking the sound and Bill was the guy that made it happen for them.

As I mentioned earlier, Bill was also the guy who did the audio post for *When We Were Kings*, the Academy Award winner, another film that quite possibly never would have been finished had it not been for the unique services he created.

Like Water for Chocolate, When We Were Kings, and Bill Markle's digital audio posting studio in the Film Center Building were three things that nobody saw coming and more than a few didn't think would ever happen. Yet they did and in a big way. When I came back from California for a visit, Bill was out of the Film Center Building and had two floors in a beautiful brownstone in the west 40s.

Regarding not seeing the potential in something, Bill tells a funny story about himself that's instructive to all of us. Back in the 60s when he was breaking into filmmaking, one of his fellow aspirants was a guy named Jim Henson (the guy who created the Muppets). Like Bill, Henson was making a name for himself working in documentaries. He was also working with puppets on the side and Bill's advice to him was, "Hey Jim, lose the puppets. They're going nowhere." Bill has a great sense of humor and appreciation of the absurdity of the human condition.

Life was good. I had just gotten the equivalent of a PhD in the practicalities of filmmaking from someone who had hands-on, black belt-level experience with every step of the filmmaking process. I took that summer off and went to Santa Fe where I immersed myself in studying Tai Chi to deal with a nagging back problem. I had money in the bank and by going out West I had also managed to avoid one of the worst heat waves the city had experienced in a long time.

In late October, after I got back, I went up to Van Cortland Park in the Bronx to play in what I thought was going to be a friendly game of touch football. It was friendly until we were one point away from winning. It was our ball and I jumped high in the air to catch the quarterback's pass and someone on the other team came under me, and jacked me up. I landed very hard on hard ground on my back. I had the vision of a cartoon piano being dropped out on the sidewalk and all the strings breaking.

I got up and sat out the last play. On the way home on the Local One train, I didn't have the energy to speak. That night my back tightened up so badly that both arms were sucked into my shoulder joints and both legs were pulled into my hip sockets. The next day, my shoulders were up by my ears and I couldn't get them down. I couldn't turn my neck left or right and I couldn't take a normal walking stride. Being able to run was out of the question.

Not being able to do these very simple physical things, which I considered fundamental to basic safety, made me very nervous about living in New York City at that time. The city was no playground then. A year after my bad fall, the number of murders in New York City peaked out at 2,245. By way of contrast, in 2017 there were only 292. Unless you lived in New York City in the 1970s and 80s, you can't imagine how bad it was.

Anway it was time to make new plans.

A few years later, while uncharacteristically watching a football game on TV, I saw a wide receiver take a similar fall straight on his back and my first thought was, "What's wrong with me that I couldn't take that?" Then I noticed he wasn't moving and he ended up leaving the field on a stretcher. "OK," I thought, "Maybe that really was a bad fall."

I wasn't going to feel more or less normal again until about 1993, five years into the future.

Chapter Three
A Competitive Edge

"To unpathed waters. To undreamed shores."

- Shakespeare, *The Winter's Tale*

The winter of 1988/89 was a hard one for me.

As a result of the body slam, the muscles in my back were perpetually locked up. No position – not standing, not lying, and definitely not sitting – was comfortable. Simple things like stepping down and up to a curb were challenging. Walking, which had previously always been a joy, was an effort, and running, something I loved and could always count on to get me out of a funk, was out of the question.

I also lost my resistance to cold. I once made it through a New York City winter with nothing more than a heavy sweater and a baseball jacket. Now cold, even just cool weather, took it out of me.

I lucked into a contract writing scripts to adapt non-fiction "how to" books into the then-new audiobook format (audio cassettes, not MP3s), so I was able to work from home at my own pace. Going to an office and fitting into the 9-5 was out of the question. I managed to adapt, but it became clear New York was not a good place to be in the condition I was in.

In addition to this, I was suffering at the time from what I call "New York City derangement syndrome." I could not imagine living anywhere else. This is a surprisingly common syndrome and back then "New York" meant Manhattan.

We didn't go to Brooklyn. Brooklyn was Siberia. There was "nothing" there. Of course, if we had families there or in New Jersey or the Bronx or wherever, we'd go there to visit them, but otherwise it was not a destination, just another place on the map that wasn't New York.

From my point of view, New York meant cheap food and abundant bookstores. It also meant a place where I could live without a car. I was and still am a serious music fan, especially of jazz and what's now called world music, and New York City was the Big Apple for both those places. Why would I even think of living anywhere else? Also, even though New York has a well-deserved reputation for being a hard place to make a living, I had the world's cheapest apartment and, before I got racked up, between tech writing, teaching speed reading and other freelance gigs I managed to dig up, I didn't have a problem paying my bills and I didn't miss too many meals.

But the winter cold was killing me, and I didn't like the idea of living in New York City and not being able to run or easily turn my head left or right and even just get my hands up in front of my face fast if I needed to. The New York City of the 1970s and 80s where I grew up was no joke and I felt like a perpetual sitting duck.

Previously, Tai Chi had helped me work out some minor back problems. Maybe it would help me work out this major one.

My teacher had moved from Santa Fe to San Francisco. I knew San Francisco. I lived there for a year when I was seven. It was 1967, the Summer of Love. I knew all about hippies and love beads. My brother and I had more than one hippie babysitter. When my mother gave birth to my sister that year, she recalled the maternity ward was filled with young women from all over the country who'd gotten a jump start on the "Summer of Love".

Things were much looser then. It was my job to take my five-year-old brother and myself to and from school every morning. We'd get on the Geary express bus at Arguello and Geary and change at Fillmore Street to go up the hill to our school. The Fillmore West was at the corner where we made our transfer and I used to puzzle at the marquee: "The Grateful Dead", "The Jefferson Airplane", and "Strawberry Alarm Clock". What was going on in there?

I didn't know it then but learned later that the Fillmore had not only been a major rock and roll center (the Fillmore West and Winterland) but also had the biggest collection of jazz clubs west of the Mississippi in the 1940s and 50s. Seeing people like Billie Holiday and Duke Ellington walking around was no big deal.

1967 was a "transitional" time for the neighborhood. The city fathers decided the land it was on was too valuable for the people who lived there and began a process of "urban renewal" which boiled down to condemning some properties for "public health" reasons, taking others by eminent domain, and then letting the commercial hold outs wither and die as the community around them disappeared. Thus a culturally rich and happily mixed-race neighborhood where black Americans owned hotels, restaurants, nightclubs, and even things like bowling alleys, was mercilessly gutted.[3]

I knew San Francisco and in the summer of 1989 decided to visit and see if my Tai Chi teacher could help me with my new problems. Except for one short visit when I was in college, I hadn't spent any time there since I was seven so I was seeing it with adult eyes for the first time – and I liked what I saw.

---

3   The neighborhood has such a rich history that thirty years later in 1997 I built what was one of the Internet's first virtual museums about it: https://www.amacord.com/fillmore/museum.

A family friend who let me stay with them during my visit recommended I go visit Clement Street which was 10 blocks of great cheap food, mostly Asian. You name it and it was on Clement Street. Not just Chinese food, but many different varieties of Chinese food and Thai food, Vietnamese food, Cambodian food, and Indonesian food. My fear that I would starve to death and/or be bored to death food-wise if I ever left Manhattan evaporated. And there were some great used bookstores, namely the epic Green Apple Books with over 150,000 books and the smaller, but well-curated Albatross II.

The next thing I discovered was that I was needed in San Francisco. My Tai Chi teacher had made a major commercial mistake. Back in Santa Fe, he was a big fish in a small pond, literally the only credible Tai Chi teacher in a New Age-y kind of place that had an insatiable desire for such things. In contrast, starting a Tai Chi school in the Bay Area was like bringing sand to the beach and his enterprise was circling the drain financially.

My experience marketing my speed reading and study classes and bringing in a quick flood of new business for my friend Bill Markle's new and unprecedented audio post-production studio gave me the confidence that I could do something about the situation. You can't build a business on flyers in the Bay Area. There isn't the same concentration of people and foot traffic as in Manhattan. But there was a directory, specifically a directory of all the alternative health practitioners in the region, which included San Francisco, Marin County, and the East Bay, primarily Berkeley.

This teacher (I've had many since, and my favorite these days is Richard Clear) was one of the first generation of Americans to go to China when it reopened to learn the internal martial arts straight from the source. It wasn't an easy task. It might seem strange now to imagine a backward China "reopening" and becoming part of the modern world, but all my life until the 1970s China was locked shut and the process of modernization was very slow.

China was pure Third World back then and it took tough Westerners to wade into that situation and find and create relationships with authentic teachers. The teacher I'm referring to did that and reached a very high level of proficiency and most importantly broke down what he learned so that it was comprehensible to modern people.

Back in the "old days," students, assuming they were lucky enough to get with a genuine teacher, were just told what to do and nothing about the theory underpinning it all. "Hold this posture." "Do this movement." And it was up to students, through hours and hours and weeks and weeks and years and years, to build a mental framework for it all on their own. Add to this the fact that the internal martial arts have a long tradition of secrecy not teaching people outside of your family or village, and definitely not teaching foreigners, these early pioneers had a hard road.

But none of that made a bit of difference to the marketing problem he had. He was an unknown white guy from out of town teaching a subject that already had a super-abundance of well-entrenched experts.

So what did I do? His wife at the time was a reporter and had shot video of him when he was training in China where the two had met. It's hard to imagine now, but this was rare and exotic stuff. I knew enough about video editing (it was still the analog days) to cut it together and make a "documentary" about him. It was just a glorified home movie, but the footage *was* rare.

Then I sat down with the big directory of alternative health practitioners in the Bay Area and cold-called them on my rotary dial telephone inviting them to the premiere of this great new film. The vast majority were pleased to get this special invitation and very often it led to a brief conversation which led to me getting their physical mailing address, which led to me sending them a reminder about the film and a description of the teacher's career, and what he had to offer.

And he did have quite a bit to offer. He was not only a high-level practitioner who studied in China, but he also had a great deal of knowledge of how to use Tai Chi and other internal practices to heal injuries and improve health. He was one of the first people to teach these subjects to the general public in the United States.

Thus in a week, this guy, who no one had ever heard of and could barely fill a class, was suddenly on the radar of every single alternative health practitioner in the Bay Area as the "go-to" guy for therapeutically oriented Tai Chi classes. Then I made some instructional videos with him which allowed him to sell to people who couldn't attend the classes. Next I found a Ponderosa-style ranch up in Mendocino county which, when his following became large enough to justify it, became the site for his annual summer retreats.

His business model hasn't really changed in any significant way since, but he did do the intelligent thing of adding some books to the mix and according to people in the know, he's now a major figure in the field, and not just in the Bay Area, but internationally.

Staying with the teacher's family while I carried out this act of financial legerdemain, I realized that I could live in the Bay Area and decided to do the previously unthinkable, leave Manhattan and my miracle apartment. If this seems a bit dramatic, many people have spent *decades* in less-than-optimal apartments in New York City, in less-than-optimal neighborhoods, living perhaps less-than-optimal lives because they had such a good deal on their rent-controlled or rent-stabilized apartments.

I was greatly aided in the venture of leaving New York and moving to the Bay Area by Bettina Mueller, my then girlfriend who I'm still with 35 years later (minus a few hiccups along the way, including one seven-year separation).

We'd met in Mexico in a little fishing village called Barra de Navidad where we were both taking a workshop with a woman named Charlotte Selver. Bettina was a friend and serious student of Charlotte's. She is also a student of Rinzai Zen, and the Japanese tea ceremony (Uresenke).

When Charlotte asked her to house-sit her home in Muir Beach in the western rural part of Marin, we packed her dented Subaru station wagon with some clothes and a few paper bags of books and headed west. I left the bulk of my books in my New York City apartment which I'd sublet to a friend. He was glad to have them. It was a great library.

It was my first and what has proven to be my only trip across the country by land. Bettina, who had lived in pre-glitz Aspen, Colorado was used to big long car trips. Me? Not so much. Previously, I had attempted to cross the country once in a Greyhound bus and 48 hours later when we rolled into Denver with only the occasional one or half-hour break along the way – no stops to sleep anywhere but the bus – my hind quarters were so sore I couldn't travel another mile. After a few days recovering in Boulder, I flew the last leg to California.

The trip across the country with Bettina was a lot more fun. A few of the stops we made along the way stick out.

On New Year's Eve, we stopped in Knoxville, Tennessee to visit my great aunt Ray (short for Raphaela) who was living in a nice nursing home under the watchful eye of my uncle, a doctor who was teaching at the University of Tennessee.

Ray, like the rest of my family, was from the Bronx via Manhattan via places like Ireland, Italy, and Germany. She was on my mother's, the Italian, side of the family, and was the sister of my grandfather.

Ray had a fantastic job when she was a young woman in the 1920s. Her boss, a dress manufacturer, would take her to places like the Cotton Club and the Onyx Club where she'd scope out what particularly stylish women were wearing, remember it all, and go back and work out the dress patterns so her employer could knock them off. She pulled off this feat of memory and visualization without formal training of any kind. I'm not sure if she finished high school, not that high school would have helped her develop these skills. The women on the Italian side of the family were phenomenal seamstresses and legend has it some of them were lacemakers to the Astors, work they did at home for which they were paid by the piece.

The men in the family were all stone masons and very good ones. When the Rockefellers bought a bunch of monastery buildings and courtyards in Europe to build the Cloisters in New York, no one on their staff knew what to do with the rocks that arrived. It turned out my grandfather, who was a partner in a granite contracting company, figured it out for them. Like my great aunt, I'm not sure he had a high school diploma. I know he didn't have a birth certificate because we hired a professional genealogist to find it and it does not exist. Things were looser in those days.

Ray also educated me about what a tenement was. She and the rest of her family grew up in one. Not on the Lower East Side. That was grand compared to the neighborhood they were in. Their tenement home was in the East 40s. Today the neighborhood is quite posh and is best known as home to the United Nations. Then it was the slaughterhouse district. Many tourists who've been to New York City are familiar with what's now called the Meatpacking District around West 14th Street. The one on the east side where my relatives lived was far larger and covered 17 acres. It was not a pleasant place to live.

Most people understand that tenements were very crowded. Families were big and units were small. Ten or more people living in two rooms was not unusual. But that's not the half of it.

The units had no electricity. So no electric lights, no fans – certainly no air conditioners – no refrigerators. And they had no indoor plumbing. In Ray's childhood home, there was one water source, a pipe in the back of the building. It might have been a pump. If you needed to wash clothes, you headed down to the backyard with pails and carried the water up. Same thing for cleaning the house, for cooking, and for casual washing. Bathing was done at public bathhouses. And toilets? Have you ever seen something called a chamber pot in an antique store? Horse-drawn wagons would go by daily and collect the pots. This was less than 120 years ago, folks. In Manhattan, the center of the universe.

I only learned these things because I used to occasionally visit Aunt Ray when she lived in an apartment in the Bronx and take her to lunch. These historical realities of life were not otherwise discussed at home or in school and the news media had total amnesia about them.

Both sides of my family moved to the Bronx which, at the time, was a paradise for people who grew up in the tenements downtown. In the 1970s and early 1980s, not so much. Ray had two arthritic knees and struggled on a cane to get up and down the stairs of her Bronx apartment. No elevator. She was on the top floor and had to run a gauntlet of young men just "hanging out." "I'm not worried about them," she said. "If they ever give me trouble, I just raise my cane." She wasn't kidding. She was old, but in her 70s, she was not frail, even if her knees hurt.

I had the occasion to go back to the place in Italy where my family is from. It's an unjustly obscure province called Piacenza, southeast of Milan and northeast of Genoa. The city of Piacenza has to be one of the most blood-soaked places on earth. It had the advantage of being at the crossroads on the route from northern Europe to Rome and the west-to-east road that connected the Mediterranean with the Adriatic. While that was handy for trade, and in 1000 AD Piacenza was one of the richest cities in Europe, it was also the place where every armed force that wanted to do something in Italy had to pass through which included the Gauls, Hannibal, the Romans, a countless number of Medieval, Renaissance, and Enlightenment-era maniacs including Napoleon, and not that long ago, Hitler.

I wrote a book about the place called *Piacenza: Hidden Jewel on the Crossroads of History*. If you're ever in northern Italy and want to get off the tourist trail and see something real, I recommend it. The food is off-the-hook great and they have a lot of specialties you won't find anywhere else.

I don't know where all my Piacenza relatives lived, but I do know my grandmother was born in Selva, in the mountains. Deep in the mountains. Three hours by car from the city of Piacenza, straight up, the last half-hour of which is a very narrow road that can only take cars and small vans. When I visited, it was late October and already sleeting. Selva was the most remote place I've ever been in my life where people lived and I could not imagine how they kept body and soul together. Clearly, they were 100% self-sufficient and tough as nails. They went from that to the tenements of the East 40s.

I was glad when we stopped in Knoxville to see my Aunt Ray on our way out west. It was the last time I saw her. In addition to being whip-smart and tough, she was one of the kindest people I've ever known, a trait she said she got from her mother, my great-grandmother, who she never stopped praising.

Ray used to play with me when I was very young. The games all had a song component to them and were pretty simple. She taught me a much-simplified version of "Wind the Bobbin Up" which I still remember. I learned later at my mother's funeral that when I was a baby I cried a lot, so much so that it was decided it was best that I go somewhere else, with my mother, of course. The two of us moved back to my grandmother's house in the Bronx and Aunt Ray, whose husband died young and who never had children, was the third of the trio of women who took care of me.

My grandmother and mother, God bless them, were not the two most stable people in the world, and I have the feeling the most important thing that ever happened in my young life was having my aunt as a caretaker when I was shiny and new and trying to make fundamental sense of the world.

Visiting aunt Ray was our first stop heading West. The next stop was New Orleans, the land of dreams.

We rolled into town very late on what must have been January 1, which means we woke up there on January 2. We didn't know anybody there and the town seemed dead, which, given the day of the year, makes sense in retrospect. We drove around a bit, saw the mansions on Saint Charles, and got back on the road. Little did I know that 15 years later this town that I didn't understand or appreciate then was going to become a huge part of my life. I wrote about it in the book *Death, Resurrection, and the Spirit of New Orleans*.

The other thing I remember about the trip was the Sonoran desert in Arizona. I don't know that I'd like to live in a desert, but I love visiting them and the Sonora is one of the prettiest deserts on earth and it was my first experience of one.

Next stop San Francisco.

We didn't have solid plans. I came out with one suitcase, Bettina likes to remind me it was a cardboard one, and two shopping bags full of books. Bettina had her stuff, as much as we could fit in a station wagon. Staying at Charlotte Selver's house for a month was great. Muir Beach is right up on bluffs overlooking the Pacific Ocean. San Francisco was only about 30 minutes away.

We looked for apartments in Marin and elsewhere to see if it was plausible to stay on after the house sitting ended. Marin was oddly expensive for housing that was not much to write home about so we started looking for places in the city, which at that time was a genuine bargain.

In 1990 San Francisco was in the doldrums. The personal computer boom had calmed down considerably. As evidence of that, a friend had a job with a research company calling CEOs in the Bay Area computer industry and asking them what they thought the next few years were going to look like. The word was not good. It seems that everybody who was going to get a computer, already had a computer, and computers were becoming low-margin commodities. This reality was reflected in the city's rental market. There were many deals to be had and we saw a lot of beautiful places.

Finally, we saw a "For Rent" sign on an apartment building on the northeast corner of Fillmore Street and California. It had a small marble lobby and an elevator, two things that were a big step up for me and Bettina. The apartment was a studio with a north and west view and big windows. We called the number on the sign, met with the landlord, he did a credit check, and we were in. It was that easy. Not that many years later, there'd be long lines of people, 10 to 20 people deep, waiting with their credit applications to get a shot at getting an apartment, any apartment.

Once we got settled, I scrounged around, looking for tech writing work. No one cared that I had top-tier banking experience; work was not easy to find. Like I said, in 1990 San Francisco was in the doldrums. San Francisco is a boom-bust kind of town. The Gold Rush was an obvious boom, followed by a long bust. World War II, boom town again, followed by a long unspectacular run, thus the cheap rents for beatniks and hippies. Then came the personal computer boom and bust.

Things got so desperate I ran an ad in the local computer industry magazine offering to pay a $1,000 finder's fee for anyone who could get me a tech writer contract. An ill-tempered alcoholic (i.e. a mean drunk) who had a contract with PG&E to create an inventory management system was being pressured by them to provide documentation of his work. I looked like the solution to his problems. He told me I wouldn't get my first paycheck until 90 days in because it was contract work. I must have been desperate because I took the job. It was easily the most unpleasant "work environment" I've ever been in.

To write technical documentation, you have to be able to ask the person who created the system how it works. He didn't like being asked questions. In fact, he didn't like talking to me at all, a point he made clear in frequent drunken tirades. Somehow he expected me to magically document the system, so I sat there and banged away and managed to make some kind of sense of it. I was able to endure 90 days of this because: 1) I really needed the money, 2) I was theoretically being paid well (and in the end he did pay me), and 3) being randomly screamed at (and worse) by unhinged people is a good summary of my upbringing (which is why moments with my Aunt Ray were so important).

I went from that nightmare to a job working as a tech writing contractor for a project at Safeway, the big supermarket chain. They were developing a system to take credit cards for payment in their supermarkets. Paying for your groceries on credit? That sounded crazy and reckless to me. Obviously, I didn't understand the direction the world was going in 1990. It was surprisingly complicated to pull this off. I remember sitting around a desk with eight people, each representing a different vendor that had to be part of the sequence that made it possible for people to stick a piece of plastic into a terminal and walk away with a shopping cart full of groceries.

The thing that I remember the most about the job was that after we were on it for weeks working productively, one day a squadron of people from Arthur Anderson arrived to make us more "efficient." We were already phenomenally efficient and these people didn't know their asses from their elbows, though they were very well dressed and groomed.

I could never figure out exactly what they did to earn their money other than make someone at Safeway HQ feel more comfortable. They spent all their time in a conference room writing on a whiteboard. In contrast, the people I worked with on Wall Street actually worked for a living. They did things. These folks were another story completely.

It was my first up close and personal experience with the white-collar grifter class, which I now understand is vast and may be the biggest employer in the United States, especially when you factor in the federal, state, and local governments.

Working for other people, especially corporate people, was starting to look like one of the 9 circles of hell in Dante's Inferno, but I didn't have a better idea at the time.

My next and what turned out to be my last ever paycheck-from-a-company job was a contract assignment with Hayes Communications, the pioneering modem maker. In those days, the online world, such as it was, ran on Hayes modems and they were minting money. Based in

Georgia, they had their R&D office in San Francisco because top technical people were not about to leave the Bay Area to go anywhere, and if you wanted access to the talent you had to be in the Bay Area.

When I was at Hayes, I wrote the manual for the company's first ISDN modem and in the process had to do a crash course – self-directed as always – on telecommunications technology. A team of five engineers had been working on this thing for a year and it was my job to figure out everything about it in a matter of weeks, write it up, lay it out, and put it all in a manual. Tech writing business as usual.

The unique spin on this job was that the engineers disagreed with each other about the details of what they had made. I went to meetings and listened to them argue about how various functions worked, and sometimes if they even worked at all. From all this, I was expected to produce a manual that made sense. They were nice guys though.

I had another "corporate" experience at Hayes which instructed me about what I could expect if I stayed on this course. I'd been on the job for a couple of months and another contractor, a woman, had been assigned to work under me. She came to the office occasionally but did most of her work at home because she had a small child. We got along fine, her work was excellent, and when childcare was an issue, and it sometimes was, my attitude was that we could work around it. All that mattered was to get the work done and ultimately done on time.

This worked perfectly well – until my boss hired a manager to manage us. Keep in mind we'd been on the job productively and congenially with the client for a couple of months by this point, but now we have a manager to report to. My thought was "OK. Crazy, but I can live with it. Just keep paying me every Friday."

The new manager, a woman, proceeded to mercilessly bust the chops of my assistant at every opportunity. For example, taking business hours time to take her kid to the doctor, which was a total non-issue to me and had no impact on her work became a huge issue to this manager. I couldn't believe it. At least the idiots from Arthur Anderson on the Safeway job had the good sense to leave us alone so we could do our work. This "manager" pointlessly undermined the morale of a capable, hard-working person who, even if she was not "on call" 9 to 5 every day, never missed a deadline, always produced first-class work, and was a pleasure to work with.

One last Hayes experience. Hayes was a privately owned company and like I said it was minting money. One day Dennis Hayes himself visited the office for our Friday, end-of-the-week employee get-together and we were all looking forward to the visit.

During the last hour on Fridays, assuming there were no fires to put out, everyone who didn't have anything pressing would get together in the employee lounge to eat, drink, and socialize. It was a very civilized thing and made sense on a lot of levels. Not much work gets done in the office late Friday afternoon, so instead of a bunch of people clock-watching, those who were inclined would relax, talk, and bond. It was a cheap perk that people looked forward to. Team building without the corporate BS. In comes Dennis Hayes. We're all standing as we typically would do at these events. He sits down...and his wife sits down on his lap... and proceeds to start feeding him grapes by hand. That was the sum total of his interaction with us.

I decided I didn't want to be a clock puncher anymore.

- Part Two -

Planting Seeds

Chapter Four
On the Launch Pad (1990-1992)

"I don't think it advisable to tell young men, or any other, to come here expecting to make their pile and return to the East. The chances of doing this, always doubtful, have nearly ceased to exist..."

– Horace Greeley (1811-1872)
updating the advice attributed to him,
"Go West, young man" after visiting San Francisco in 1859.

San Francisco, post-personal computer boom and pre-Internet, was a bleak place to be looking for work.

The good news was that rents and lots of great places to eat were cheap. The great societal "speed up" that commenced with the ascendance of the Internet, with San Francisco as Ground Zero, hadn't yet started.

Being gainfully unemployed, Bettina and I had a lot of time to hike around Marin, make occasional road trips further north to Mendocino and Humboldt, and explore the city. I spent hours every day in the city's bookstores and the gracious Old Main Library.[4]

San Francisco was a great bookstore town. An article from the 1996 San Francisco Chronicle said there were over 100,000 square feet of bookstore space in the Union

---

4   In 1995, the library was closed and replaced with a "new and improved" building – minus card catalogs and with 33% less shelf space to make room for computer stations and grand architectural flourishes. The library handled the shortfall of shelf space by surreptitiously shipping the "surplus" books to the landfill. An ominous forerunner of digital things to come. Novelist Nicholson Baker blew the whistle on this scandal in an article in the New Yorker.

Square area alone and small, quirky bookstores seemed to be shoehorned all over the city in every neighborhood. We had three within five minutes or less from our apartment. Stacey's on Market Street downtown had a great selection of business books.

My family lived in San Francisco for one year in 1967 when I was seven. It's where I got my first library card which I still have nearly sixty years later. In terms of the practical impact it had on me then and still has on my feelings now, it may be the most prized document I have in my archive.

If you offered me an original copy of the
Declaration of Independence in exchange for this
old library card, I'd keep my card.

First, it had my name on it, a big deal for a seven-year-old. Second, it allowed me to go to the library and take home any books I wanted. It may have been my first conscious experience with abundance and the very existence of libraries still amazes me. The third thing I liked about the card was its expiration date: 1970. Not only did that seem impossibly far away, but it had a very space-age feel to it.

I tried to imagine what life would be like in 1970. As soon as I learned how to put a letter together - letter, stamp, envelope - I got the address of NASA and wrote them a bunch of questions. Much to my delight and long after I'd forgotten I'd sent the letter, I got a reply, of sorts. No answers to my questions one of which was "After we get to the moon, what's next? Mars?", but they did send me a bunch of color brochures.

To this day, I write letters all the time to people I don't know, but whose work has impressed me. I get a response rate of about 1 out of 100. On one of those occasions, a letter began a relationship that led to me making several million dollars. On at least two other occasions that I'm aware of, my reaching out by letter to people at the beginning of their careers led to *them* making millions of dollars.

My brother, age 5, and I were free-range kids to the extreme. I'd take him to school in the morning by public bus and when we'd get back, we'd change out of our school uniforms and run to the park where we'd stay until dinner. San Francisco has forgiving weather which means when we were not in school or not home at night, we were outdoors. Our favorite place was the Presidio, now a big park and then a U.S. Army base with a big park around it, and what was then called Julius Kahn Playground.

The playground has since been redesigned "with the feel and aesthetics of the Luxembourg Gardens in Paris". Then it was just a fun place to play. There was a big hole in the chain link fence surrounding the park that let you go into the woods of the Presidio and we were back there often.[5]

---

5   I learned later that in 1969 that park and the hole in the fence was the last place the Zodiac killer was ever seen. He was escaping a dragnet that was chasing him down after one of his murders. For all its postcard beauty, San Francisco has a darker-than-average edge to it.

San Francisco was also where I started writing. In second grade, we were given an assignment to look at a picture and write a story about it. Most everyone wrote one or two lines of description. I took the assignment literally and wrote a story, a full page worth. My teacher was astonished and so were my parents. I was surprised everyone couldn't do that. It seemed normal to me, but they thought it was a big deal, and I, not looking a gift horse in the mouth, enjoyed the positive attention.

A few years later, I made the decision I wanted to be a writer. So I wrote. Starting in 6th grade, I had the self-imposed discipline of writing ten pages long hand every day, about anything, stream of consciousness style. Later, in high school, an English teacher pointed out that good writing has a definite beginning, middle, and end. It was a revelation to me and is the only writing lesson from school that I remember.

As an adult, being back in San Francisco was like being a salmon swimming upstream to return to where he was spawned. I was back in the spot where I learned to use a dictionary, write a letter, take books out of the library, and where I started writing, but none of these things led to making a living – yet. Bettina and I had leisure but not much money and the money we did have was slowly and steadily melting away.

I could have polished my resume and dove back into the tech writing cesspool, but the idea of going back seemed like a fate if not worse than death then at least close to it. I could have started up my speed reading and study skills courses again, but I'd been there and done that and didn't have much enthusiasm for the idea.

I did manage to land a contract from Prentice-Hall to write a book for their "Made Simple" series, in this case, *Speed Reading Made Simple*. The deal was $2,500: $1,250 now, and $1,250 on delivery. It was a thrill. The money came in handy and I was going to be a published author. After I sent them the first draft, they wrote back and said they wanted the book to conform to their other titles with lots of filler questions and answers and other nonsense. I had spent hundreds of hours successfully teaching this material to a few thousand students and I wrote the book exactly the way I judged it needed to be written to achieve its goal. We agreed to disagree and I sent the advance back. Ouch. But I still have the manuscript which I will self-publish someday, probably in 2025.

One night, I was out on my beloved Clement Street (good, cheap food and great bookstores) going through the stacks in the bookstores. Browsing books, whether in bookshops or libraries, is a great form of free entertainment and I spent countless hours doing this.

I walked over to Albatross II Books, a small store, much smaller than the vast Green Apple Books on the same street, and started going through their discount bin which was in the middle of the store. An 8 1/2" x 11" paperback with a red cover in terrible shape, the binding was coming undone, jumped out at me. It was called *Mail Order Know-How* and it was by Cecil Hoge, Sr. It was all about mail order which is selling at a distance vs. selling face-to-face on a sales call or in a retail store. If "mail order" sounds like a quaint, old-fashioned term, Amazon, which was still years away from conception is, in fact, a mail-order business. The only difference between it and an old-school mail-order catalog is you place your order online vs. sending a check and order form.

The book was fascinating, engaging, and filled with stories of people taking little more than ideas and building big, sometimes very big businesses around them. This was just the kind of thing I was looking for and I realized that unbeknownst to me I had already been in the mail order business, or as it came to later be called, direct marketing.

The "direct" in direct marketing means you sell directly to your customers. You don't do it through distributors or wholesalers and your "store" is essentially a warehouse and a shipping department. This means it can be anywhere and you can start very small and, if successful ramp up as needed. Instead of a store window, you run ads and people contact you directly.[6]

Mail Order Know-How was filled with eye-catching chapter titles like "How to Raise a Billion Dollars" which told stories of people who started successful companies with little more than a shoestring. The author of the book's *To the Reader* section was introduced as someone who "started as a moonlighter with a one-inch ad, and, via mail order, built a Fortune 500 company - and then in his early forties, sold out, and retired." That sounded pretty good to me.

Despite how appealing the book appeared and how in line it was with what I needed, I was so broke at the time that the $10 price tag gave me pause. It gave me so much pause that I put it down and started to look at other books on the shelves. But something drove me back. I picked it up again, flipped through it, and then put it down again. I repeated this ritual at least two more times. Finally, I said "F*** it. So I'll miss a few meals. I'm already doing that, so what's the difference?"

I brought my prize home and for the next year, my mind was aflame. First and foremost, I realized that what I had been doing by instinct was a thing that people, lots of people, did for a living. There was a whole industry built on it. It was called direct marketing. All the details that I sweated over - headlines, response devices, and what I later learned was "ad copy" - were things other people were

---

6   My friend Steve O'Keefe, the man Jeff Bezos consulted with when he was thinking about going into the business of selling books online, said when he visited Amazon's original "headquarters" it was a small and shabby office suite in a bad part of town. The staff, I believe it was five of them then, used doors with sawhorses as their desks.

sweating over too and, most importantly, writing about which meant instead of re-inventing the wheel, I could learn from others.

Hoge's book focused on the stories of people and their businesses and careers and every time I read about a new person I'd check to see if they had any books. Very often they did and I'd get their books and devour them. Then I discovered that there was a thing called *Direct Marketing Magazine*. It was filled with case studies and articles by practitioners. The magazine also held conferences and sold recordings of sessions (audio cassettes in those days). My head was exploding, in a good way. I spent an entire year doing pretty much nothing else but reading and listening to tapes on direct marketing. At one point Bettina asked me where the heck this was leading. I did not know, but I was absolutely fascinated by what I was learning.[7]

Some of the people who especially impressed me were David Ogilvy, John Caples, Victor Schwab, Richard Benson, Maxwell Sackheim, and Eugene Schwartz. Luckily, the public library had all their books and many more on mail order and direct marketing.

Luck did play a role in this because, as I mentioned earlier in this book, in 1994 the main branch of San Francisco's public library surreptitiously sent nearly a third of its books to the landfill because their new building put more emphasis on grand atriums and computer workstations than bookshelves.[8]

---

7   Many decades later I was comparing notes with Richard Viguerie, who is universally acknowledged as the pioneer of political direct mail, and he related a similar story. A young man discovers direct mail which, like Internet marketing is a subset of direct marketing, and his head explodes. He too spent a solid year doing nothing but studying the field, but had the good sense to negotiate the process up front with his wife.

8   A few years later when I went to check out Maxwell Sackheim's *My First 65 Years in Advertising* and Eugene Schwartz's *Breakthrough Advertising*, two essential masterpieces in the field, they had been sent to the landfill and were no longer on the shelves.

Another book that was very important to me was Drayton Bird's *Common Sense Direct Marketing* which was new then and went on to become a primary textbook of direct marketing, especially in Europe and Asia (Drayton is based in the UK).

More than any other book I'd read up to that time, it brought the practice of direct marketing down to earth and gave me the *feeling* that "Hey, I can do this." All information in the world is not of much use if you can't imagine yourself in the game and Drayton's book got me over that all-important hump. I found the book in the main branch of the New York Public Library on a trip back home.

I had the books, the articles, and the tapes, but I was lacking two things: 1) living, breathing peers, and 2) a real-world project to apply what I was learning.

Finding peers took a while, but the vehicle came along in the form of declining interest rates. The average rate for a mortgage in the U.S. peaked at 10.13% in 1990 - the year I arrived in San Francisco - and then began a slow and steady decline. By hanging around the investment banks in New York as a tech writer, I learned that when interest rates change, the world changes with it. This manifested in real estate in two ways:

1. If you took out a mortgage at say 10%, it made perfect sense to re-finance it when rates went lower. Why? Because your monthly payment would be less. Yes, you might have some fees to make the change, but with third-grade arithmetic, you could paper and pencil it out and see if it made financial sense. More often than not, it did, especially as rates crashed from 10% to 9% to 8% to 7%.

This greatly expanded the ranks of what had been a relatively obscure profession, the mortgage broker. These are people who, for a fee, would help you find a lender, negotiate your rate, and then handle all the paperwork. All you had to do was provide your credit rating, and your mortgage payment record, and sign a few papers. Done. The broker would make around $1,000 for his or her work.

2. High interest rates were only part of the problem for the rate-sensitive real estate market. With the uncertainty the high inflation rate caused, lending standards became very strict and it was occasionally impossible to get a loan at any rate, so people who had real estate they needed to sell occasionally faced the prospect of no buyers. Why? Because very few people can pay all cash for a property. A mortgage is almost always needed to close a deal.

So what were sellers to do when buyers could not get loans? If the owner owned the property free and clear with no loans on it and 100% (or close to it) equity, the seller could become "the bank". The seller would take a down payment and then "carry" the rest of the purchase via a "seller carry-back" mortgage. Just like a bank, he'd receive monthly payments from the buyer. In states like California, Florida, and Texas which use deeds of trust, instead of traditional mortgages, lots and lots of this "paper" was created. [9]

As an example, a typical seller might be a retired couple that bought their property in the 1960s for $15,000 and it was all paid off. Now in the early 1990s, it was worth $100,000 or more and no one could get a loan to buy it. At this point in their lives, the sellers needed to sell, so they took a down payment of say $20,000 in cash and the balance of $80,000 in a note. Not ideal, but the house got sold, they got more immediate cash out of it than they put in, and

---

[9]   Deeds of trust and mortgages function the same way. They are loans secured by real estate and the buyer makes monthly payments to "the bank" to pay off the balance. The important difference is what happens in the case of default when the buyer stops paying. In mortgage states, like New York, the process is a legal nightmare and requires a court case, always a slow, expensive, and aggravating process. A judicial foreclosure. In the deeds of trust states, the process is much simpler. The buyer gets a reasonable time to catch up on delinquent payments and then boom, the property goes up for public auction to make good on the balance of the loan. Non-judicial. Fast, easy, and cheap.

they'd be receiving monthly payments from the seller until the loan was paid off or the buyer re-financed as many did.

A small, but lucrative marketplace developed made up of people who dealt in the buying and selling of owner carry-back mortgages. Yes, the owner of a mortgage - the one *receiving* the payments - can sell the mortgage to someone else, and that buyer will then start receiving the payments.

It's very similar to buying a bond with one very important difference: the bond market is liquid which means there are lots of ready buyers, whereas the market for seller carry-back paper is very specialized. This means that buyers could often get amazing discounts on face value, creating eye-popping returns of 20%, 30%, or more. If all this sounds complicated, it's just Finance 101, which thanks to my time at First Boston I had an A+ in.

How did this lead me to make money and join the world of people who knew they'd always have enough money for groceries and rent? First, I took courses in mortgage brokering and its esoteric cousin, mortgage investing. This led me to attend my first conference on the subject which I went to with high hopes. But as the conference dragged on and I met the people actually in the business and learned what the day-to-day of the work actually was I got the sinking feeling that I'd just wasted several months of my life. No matter how lucrative the business was, and it was quite lucrative for the small shops who knew how to work it, it was when all was said and done, boring.

The fact of the matter is that many, many, many highly lucrative businesses and professions are deadly boring. Making money is not about you being entertained. It's about making money which at the end of the day is profitably providing a standardized product or service to the largest number of people possible at the highest profit margins possible. It's creating a machine and then showing up every day to turn the wheel.

For better or worse, that's not me. I need new wheels to build, ideally in widely disparate fields and, as often as possible, in fields that are new to me - and I need them on a regular basis. Not a recipe for making money, but eventually I did manage to make money too.

In the meantime, I had to figure out a way to keep food on the table without the soul-crushing necessity of working for someone else. What to do? Then it dawned on me that even though these mortgage folks were doing well, they all wanted to do better and the path to doing that was doing more deals, and the path to more deals was better marketing – and they knew next to nothing about marketing.

The vast majority of their deals came as the result of networks they'd developed with fellow mortgage brokers and other real estate industry people. The idea of actively reaching out directly to the public to find deals was only barely on their radar and few had a clue how to do it, but they were all very interested in learning.

The guy who I'd taken mortgage investing classes with had a newsletter and I offered to write a monthly column on the subject of how to find more deals. I just wanted a four-line ad at the end of every article. He liked the idea and I came up with a primitive offer. For $1 mailed to my address, I'd send the mailer a free report "Five unusual ways to find mortgage deals" (or something like that). It was my old speed reading course model. Run ads and give away something free to people who inquire.

By doing this I got the name and mailing addresses of people who were in the market for what I had to sell. This gave me the opportunity to demonstrate my expertise with the material that I sent them. When that first $1 arrived, I felt like the first person who rubbed two sticks together and started a fire. Every month, more dollar bills arrived, and more importantly I was assembling a list of people who had the means and motive to give me a lot more money to learn what I knew.

This is the first dollar that came to me from my first mail-order ad. It's framed in a prominent place in my office.

But what exactly did I know? I had built a business from scratch that gave speed reading classes. I'd put a digital post-production studio with no customers on track from a standing start to profitable in just 30 days. With little more than string and glue, I saved the bacon of a Tai Chi master who'd moved his family into one of the most expensive housing markets in the world and had no customers for his classes. And I'd read mountains of books and articles about marketing and advertising.

But did I really know anything? I had this nagging feeling that since I'd never taken a class in marketing or advertising and had no credentials, I was stretching it to present myself as the solution to the mortgage industry's marketing problems. Then I remembered a saying direct mail copywriter John Tighe is credited with coining: "In the land of the blind, the one-eyed man is king." Thus Marketing Solutions for Mortgage Investors was born.

The mortgage newsletter I had volunteered to write articles for was putting on a conference and I suggested

I give a talk on marketing. Thanks to three months of columns, I was already a semi-celebrity in the market and the publisher said yes.

I'd been looking into the business of public speaking and one of the stars of that field then was a woman named Dottie Walters who'd written a book called *Speak and Grow Rich*. It was starting to dawn on me that if I could get myself in front of relevant sources as an educator that would be better than any advertising I could buy. Instead of paying for an ad with only a fraction of a fraction of likely prospects who'd see it, I could have the undivided attention of a group of people 100% of whom had the means and motive to buy what I was selling. Not only that, I could get in front of the audience for free, not counting the cost of getting there.

In the back of Walter's book on the speaking business, she put her phone number with an offer for something or other. I called the number and was surprised it was the great woman herself who answered the phone. I don't recall exactly, but I think that the offer I was calling to get was a free report.

After the pleasantries, she proceeded to start selling me what seemed at the time to be a very expensive home study course. I demurred. She correctly pointed out that if I wanted to make progress in the field, I needed an education. I didn't budge at which point she told me about her newsletter. The price was in the ballpark and I said yes to that.

I'd just experienced a very professional upsell (the home study course offer) and a down-sell (the less expensive newsletter offer). Walters made some money from a 10-minute phone call that cost her zero in ad dollars to generate and I got what turned out to be the greatest bargain of my life.

The newsletter arrived sealed in plastic. It was glossy and four-color and didn't really grab me, but there was an insert that did. It was printed on both sides on plain paper in what I'm guessing was size 6 font. It might have been even smaller. The design was so plain that it stood out. The content stood out even more.

The insert made the case that a collection of mail-order ads for books and home study courses (info products) that the author Dan Kennedy had assembled was so instructive and full of profitable examples to learn from that it was worth $95. A few months earlier, I wasn't sure I could afford a $10 book, but now that I was gearing up to sell marketing training to mortgage people, I needed all the ideas and examples I could get so I bought it. When the book arrived, it was eye-opening and liberating.

Up to that point, most of the reading I'd done about marketing, as informative and inspiring as it was, used big companies as examples. David Ogilvy's clients were Shell Oil, American Express, and Lever Brothers. Richard Benson courted big mailers like Time-Life, Hearst, US News and World Report, Children's Television Network, and Sunset Magazine. Maxwell Sackheim was famous for co-founding the *Book of the Month Club*. Other authors I learned much from - Ed Nash, Bob Stone, Freeman Gosden, Lester Wunderman, Denny Hatch and Joan Throckmorton - all did their work for big corporate clients. As a one-man shop, starting on my kitchen table, it was hard to relate.

Though he occasionally did business with big corporate clients, Dan Kennedy mostly avoided them and focused on small operators. In the same way that I was aspiring to become the go-to marketing advisor to the mortgage industry, something a lean operation can realistically tackle, he positioned himself as the go-to marketing advisor for a number of industries: professional speakers, dentists, and chiropractors – all professions that need to and are willing to spend money on marketing, but have few, if any, credible sources for advice customized to them.

He also taught clients how to make themselves into the go-to marketing advisor in a wide variety of other niches: carpet cleaning, financial planning, music schools, martial arts schools, auto repair, restaurants, HVAC, and many other "bread and butter" businesses that, when managed and marketed skillfully, have the potential to spin off large amounts of cash for their owners. Not only did the clients he taught to be industry leaders profit, but their clients thrived as well because, for the first time in their lives, they were getting marketing advice that was rooted in experience and reality, not academic theory, ad rep hyperbole, or wild and uninformed guesses.

It's possible, with time, to become a millionaire with any one of these bread-and-butter businesses, especially if you have multiple locations, or can scale otherwise. Between Dan's direct clients and the clients of his clients, many people ultimately found themselves with net worths way beyond their previous wildest imaginings. I was to become one of them.

But first, I had to do more than give away a free report for $1 shipping and handling, so I worked on my presentation and followed the advice of Dottie Walters and Dan to produce something to sell from the stage. My first serious product was a 67-page report called *Testing: The Key to Direct Marketing Profits* which I published in 1991. (Now available on Amazon.) I wrote it for two reasons: 1) I needed a product to sell at the conference and 2) I wished that such a book existed. Almost every book on direct marketing talks about testing, but I couldn't find a source that went as deep as I would have liked, so I created my own.

Testing is the key to direct marketing and it's what differentiates it from the kind of image ads that ad agencies and other uninformed parties run. Direct marketing has a few essential principles: 1) it measures inputs and outputs precisely, 2) it adheres to the philosophy that "no one knows, let's test it," and 3) it's constantly testing marketing elements to find ways to improve the return on ad dollars spent.

The simplest form of testing is called A/B split testing. Here's a simplified version of how it works in direct mail. You create two very different ads and send one ad to half of your list and the other ad to the other half of your list. You "key" the ad so that when people respond it's clear which ad they were responding to, the A or the B ad. Then you count heads. Rather than imagine or debate which ad generates the most response, you just look at the response and that's your winning ad. As soon as you have a winner, you set up another A/B split test and see if you can replace it with an even more response-generating ad. Sometimes you can. Sometimes you can't. The important thing is to be constantly trying.

Google engineers ran their first A/B in 2000, February 27, 2000, to be exact. By 2011, they ran over 7,000 tests in one year. That and recognizing the relative value of clicks is the foundation of Google's business. Take those two elements away, especially selling clicks, and there is no more Google, or Facebook, Instagram, Twitter, or TikTok for that matter.

Anyway, I printed up 25 copies of *Testing: The Key to Direct Marketing Profits* with the idea that I might sell some of them at the conference after my talk. Anything I didn't sell, I could take home and figure out what to do with later. I needn't have worried. At the end of my talk, I mentioned the book and that it was available for $20. 10 minutes later, I had no more books to sell and $500 in twenty-dollar bills in my pocket. At the time, I had no way to take credit card payments and people still carried cash.

Bettina and I celebrated with a lavish lunch at a high-end Chinese restaurant overlooking the East Bay just down the road from where the conference was. We weren't going out to eat much then – actually, we never went out to eat at that time – so it was a very special occasion. "I think this is going to work." I had stumbled on a market that had an ongoing, compelling need; I was their only source; and the value proposition for them was very good. Just one extra plain vanilla mortgage deal was worth $1,000, and one good seller-carry-back deal could easily be worth five to ten times that amount. In effect, I was selling ten-dollar bills for a dollar.

The first workshop I gave, which I filled by sending mailings to the list I'd gathered from the articles I wrote for the industry newsletter, was a day long and featured me talking about copywriting (ad writing), the owner of the letter shop that I used to get my letters into the mail, and an expert on using the phone effectively with the prospects who called in. My speed reading classes back in New York would have about 10 people per class at $195 a head. I started with this group at $295 and twenty people showed up to the first one.

Because the idea of a course on marketing designed specifically for mortgage people was so unique at the time, that first workshop was attended by many of the industry's high profile, heavy-hitters. They were leaders for a reason. They were always looking for an edge. Fortunately for me, they saw tremendous value in what I was teaching and were willing to personally endorse it to others, which was gold in a small tight-knit industry.

As time went on, I expanded the trainings to include industry people who had their own unique perspectives on the marketing side of the business. Within the space of a few years, between live attendees and tape buyers, we'd sell 50 to 100 units at between $295 to $495 per event and were doing several events per year.

There were costs of course: the mailings, the meeting room rental, someone to record the audio, duplicating the workbooks, producing the audio cassette, but the margins were great. There was one other element: sweat equity. It was not enough to just tell people, "Hey, we're doing a new workshop." I had to organize original content and write compelling sales letters to drive people to the event.

I'd been teaching myself to be a copywriter and managed to learn how to do the most important thing a copywriter does: position the offer, in this case, the workshop, as the most value-laden epic event imaginable, unlike anything ever seen before or likely ever to be seen again. I don't think the concept of FOMO (fear of missing out) had been coined yet, but I was a master of it.

Every event and thus every letter was different and every event meant a big mailing and mailings were expensive: printing, handling, and especially postage. It wasn't just a matter of pressing "send" on an e-mail or a tweet. If I failed to make the case for a given event powerfully enough, I'd be looking at a bunch of empty seats and facing a loss for all my work and investment.

Every message to my audience cost me thousands of dollars upfront so if I didn't get it right, instead of a gain, rent and food would be an issue again. Luckily that never happened. As a positive side effect, the process focused my mind considerably and I became a better-than-average copywriter.

And most importantly in the short term, the problem of making money without punching a clock at a job was over, and it turned out to be over for good.

Chapter Five
The Phantom Phone Booth

"Ancora imparo…" (I'm still learning)
— attributed to Michelangelo at the age of 87

If at the beginning of 1993, you were to ask me "What do you know about the Internet?", I could have answered you in two words: "Not much." And I cared even less.

I'd actually "seen" the Internet in the late 1970s when I was in college, but I just thought of it as a "pipe" that connected the mainframes at Princeton to those at Columbia University. I didn't know it had a name at the time which was ARPANET.

When I first got to San Francisco, in 1990, one of my employers wanted me to get a modem for my home computer. All that meant to me was that she'd be able to send me work at home which did not sound like a good idea. No thanks. I procrastinated long enough so that she forgot about it and that was the end of that.

I vaguely remembered a *New York Times* story about the Morris Worm that froze 6,000 computers on "the Internet". Back then, the Internet had a grand total of 60,000 computers on it, so I'm not sure what the big deal was. It might have had something to do with the fact that the virus' author was the son of one of the NSA's top data security experts.

The next time the Internet made it to the front pages was in 1991 when it played a role in reversing a coup in Russia (August 19-21) in which communist hardliners briefly seized power from the the newly elected president

Boris Yeltsin. Russians knew something was up when they woke up on August 19 to see the state TV broadcasting a recorded performance of Tchaikovsky's *Swan Lake* on a loop. What they weren't seeing were the tanks and troops rolling into Moscow to seize control of the country. The Internet was one of the workarounds that helped people get messages to and from Russia and around Russia when most other channels of communication were shut down.

The next time I heard about the Internet, I was visiting a friend in Humboldt County. He had a small office/cabin behind his house and he invited me to come see something in it. He fired up his computer and on the screen was the catalog of all the books in the Library of Congress. He was thrilled. "I'm looking at the card catalog of the Library of Congress!" "OK," I said. "I'm doing it via the Internet." I didn't get the significance of what he was showing me.

But fate conspired to change all that.

A few months later, a cookbook Bettina wrote was nominated for the Julia Child Cookbook Awards in the vegetarian category. It was called *A Taste of Heaven and Earth* and thirty-plus years later it's still a cult classic. Whenever I come across copies of it at yard sales, it's invariable well-used, often with food stains on it, the sign of a good cookbook.

How the book came to be and what happened after it came out is what propelled me into the world of the Internet.

When we first arrived in San Francisco, Bettina worked as the cook for a three-week retreat in Napa County and managed to keep the attendees happily fed for $3.62 per person per day. The cost of things was much lower then, but this was still an impressive feat of culinary economy.

A few years previously, she'd been co-owner of *The Beat'n' Path* in Hoboken when Hoboken was an old-school, mixed Italian-Cuban neighborhood, and still very much an urban frontier to the people who worked in Manhattan. The restaurant started as a macrobiotic collective, then quickly became a two-owner vegetarian restaurant, then it added meat to its menu, then it added a bar, and finally, it turned into a nightclub. It was in its last incarnation that it actually made money and when Bettina walked away from the business she owned an apartment near the PATH train, free and clear.

She had cut her teeth as a cook in a most unusual way. While visiting her mother in the Hudson Valley, she went to work for a guy who was making money cutting up and recycling big decommissioned diesel fuel tanks that had then lined the banks of the Hudson River. Through that job, she met a tugboat captain who was hauling empty sand barges from New York to Buffalo via the Erie Canal. He needed a cook for the crew and after convincing a female friend to join her, Bettina signed on for the three-week eight knots-an-hour trip north and then west.[10]

That cook's job led to Bettina becoming a line cook at the *East West*, one of the pioneering macrobiotic restaurants in New York City which counted John Lennon and Yoko Ono as customers. Between the tugboat, cooking at *East West*, and running the *Beat'n' Path*, she knew a thing or two about feeding people and the retreat was a chance to exercise those muscles again and re-visit some of the dishes she'd perfected at the *Beat'n' Path*.

I'd been working on my speeding reading book and Bettina, being the ever-industrious person that she is, when she got back from the retreat, she decided that she'd figure

---

10   Bettina's friend Pam Hepburn took to the experience so much she ended up becoming a tugboat captain, one of the first and now the longest-running woman tugboat captains in New York harbor. It's a very serious job. When the Twin Towers collapsed on 9/11 Pam was the person who figured out how to dock the retired NYC fireboat, the John J. Harvey, at the dock-less shore so it could pump water from the river to firemen to suppress dust, clean them off, and put out countless small fires. For four days it was their only source of water at the site, the collapsing building having severed all the water mains.

out how to use my computer (I was the first person she knew who owned one) and write a cookbook based on the meals that she'd made for the retreat goers.

Later at a dinner party when a friend asked her what she was up to, she told her, "I'm working on a cookbook." To which her friend replied, "I have a friend who is a book agent. I'll tell him what you're up to." One thing led to another and, in a shockingly short period of time, Bettina had a $30,000 advance to write a cookbook. It turned out that the friend was a book agent with an extremely hot hand, knew and admired the person who led the retreat Bettina had cooked for, and wanted to see more in print about her which the book promised to do. Bettina had shown up at the right place at the right time with the right marketable idea. It was that easy.

But there was one catch. She wasn't a writer. Other than high school papers, postcards, and personal journals, she hadn't done much writing and had never written with an audience in mind. The agent and publisher Harper Collins told her, "Get a writer and pay them out of the advance." She didn't have to look far for a writer. She was living with one. Bettina wrote down the recipes she'd created (and tested them on me to make sure they worked as instructed) and I wrote the book's connecting essays based on themes she wanted to cover.

The next thing we knew we found ourselves with a nomination for the Julia Child Cookbook Award. The rules worked as follows: 1) winners in specific categories would be selected by a committee of relevant content experts and 2) the number one prize for best cookbook of the year would be up to a popular vote.

I had my eyes on the latter prize and sent ballots to every living person I knew asking them to vote for Bettina's book. I left no stone unturned, but there was one person I just could not find, David Madole, a college schoolmate, who was a composer with a lot of experience with computers and electronic music. After much searching, I found someone who had his e-mail address.

One problem. Not only did I not have an e-mail account, I wasn't even sure what an e-mail address was. Thinking my friend up in Humboldt might have an idea what to do with this thing, I called him and asked if he could send an e-mail to David with my phone number. He did and a few days later, the phone rang. It was David and amazingly, he was living just five blocks from my apartment in San Francisco. I say "amazing" because the last I knew he'd moved from Princeton where he was a grad student to Florida State University where he taught composing. He'd just arrived from back east and in less than a month he was moving to Oakland near Mills College where he was going to be part of the electronic music program there.

This was important, not just because he voted for our cookbook, but because he was the first serious, living and breathing "Internet user" I had ever met. He'd been one of the very first people at Princeton to have an Internet account back in the 1970s. I asked him what this e-mail and Internet thing were and he gave an explanation which must have made an impression on me. "You should look into it." I wanted to know more, but it was clear he was not inclined to be my tutor and it was going to be up to me to figure it out. Fair enough.

As we were parting, he suggested a magazine for me to get, a quirky, niche publication called *Boardwatch*. It was not a magazine that normally would have been carried by the chic magazine store on Fillmore Street midway between where David and I lived. But it was. The publisher in a fit of grand ambition had massively overprinted and his distributor somehow persuaded the store to carry it and make room for it on the shelves. I had never seen that magazine anywhere before and I never saw it on a magazine rack again.

But there it was exactly when I needed it and the issue was promoting the upcoming annual *Boardwatch* convention coming up in six weeks in Colorado Springs. I read the prospectus and it was clear with its 100+ information sessions, trade show floor, and a few thousand online enthusiasts from around the world, that this was the place to go if I wanted to find out what this "online" thing was. Not three years previously, I had sweated over the speculative purchase of a $10 book. Now I was looking at about $1,000 to get to and fro, stay at the hotel, and pay the conference fee.

Was it really going to be worth the hassle and expense to fly out to Colorado? I wasn't sure. After all, I couldn't see any evidence from the world around me that the online world was anything but a hobbyist's playground, and I was living in San Francisco, then the most digital place on earth. My business was keeping me plenty busy. I already got the big lesson that I needed to get online somehow and get an e-mail address, but even that seemed to me to be a low priority. Still, I was intrigued and like that $10 book in the remnant bin that I almost didn't buy, I kept rolling it around in my mind. "You should go," Bettina said. "You have the money." Yes, now I did, but $1,000 seemed like a lot to spend on something that appeared to have no relationship to what was then my core business. I was torn.

Then I had a dream...

It was one of those very realistic dreams. I was in a plausible-looking version of San Fransisco. I was walking downhill from Van Ness Avenue to Polk Street, a walk I'd taken scores of times. Very plausible. Before I got to Polk there was a phone booth. Nothing out of the ordinary there, but this one was made of lacquered oak. A little strange, but not enough to tip me off that I was dreaming. In the phone booth was a drawer with a polished brass handle. I pulled it open and it was filled with rolls of hundred-dollar bills wrapped in rubber bands.

When I woke up I knew exactly what the dream meant: There was a lot of money to be made in "telecommunications". I booked my flight and ordered my conference tickets - by phone - because it was 1993 and there was no way to buy conference tickets or anything else online.

Colorado, here I come.

Chapter Six
"This Machine is a Server"

"This machine is a server. DO NOT POWER IT DOWN!!"
- A label with red lettering that Tim Berners-Lee posted
on a computer in his office, which was home to
the first web server. c. late 1990

Using a back-of-the-envelope estimate, roughly 4 billion people have been born in the 34 years since 1990. Of those souls who are still with us, they have all grown up in a world with a commercial, public Internet and many, especially in the developed world, have no memory or experience of what the world was like without it.

The Internet has not changed basic human nature or power politics in which the strong attempt to permanently exploit the weak in ever-escalating ways, but it has, without exaggeration, changed every detail of everyday modern life: how people learn and get education, how they get their news, how they work and make money, how they entertain themselves, how they search for information, how their workplaces are managed, how they find friends and mates, how they connect with people outside their communities and workplaces, how products and services are sold to them, and more. The operational details of every human activity and institution, without exception, have been disrupted and some so profoundly as to make them all but unrecognizable.

The change was not instant and it was, and still is, not universal, but, as of this writing (2024), 5.45 billion people, roughly two-thirds of the world's population, and nearly

100% of the population of the developed world have some kind of access to the Internet.

This was not the case in 1993 when I attended ONE BBSCON. In late 1993, there were only 500 known web servers worldwide and only 623 websites. The Web, which is how most people access the Internet today, accounted for only 1% of all Internet traffic. Of the 5.557 billion people living then, only about 14 million were on the Internet. That's 0.025% of the world's population. In other words, darn near nobody. And most of those who were "on the Internet" were "on" it only in the sense they could send and receive Internet e-mail which became ubiquitous for online users in 1993. Other than people using the e-mail system, the vast majority of people using the Internet in 1993 were scientists, academics, military people, and very serious hobbyists.

In contrast, for most people being "online" had nothing to do with the Internet. It meant being a user of computer bulletin board services (BBS). Computer bulletin boards can be broken into two basic categories: big ones and little ones. You could count the big ones on the fingers of one hand: CompuServe, Prodigy, GEnie, Delphi, and AOL. In addition to these, there were approximately 70,0000 known computer bulletin boards which ranged in size and sophistication from the WELL to computers with a single modem attached to them.

Big or small, all of these services were reached via phone landlines only. (Cell phones were extremely rare then.) The user's computer had to have a modem, and would reach the service of the user's choice by dialing a specific phone number. If there were an available modem on the BBS side, you'd hear the connection being made,[11] and if all went well you'd be "online," but only on the one specific computer bulletin board service you were logged onto.

---

11   This is what it sounded like: https://bit.ly/modem-sound

There was no streaming audio, no video, and not even pictures on pages then. Images were a big deal on BBSs but you had to go to a special section and download the images you wanted to look at individually. BBSs were a text-only experience, and operated by a command-line interface. No point and click. In contrast, the Internet was a rare bird that at the time most online users didn't know anything about.

Further, if you wanted to access another computer bulletin board, you'd have to disconnect from the one you were on and dial up another one. Each service was its own island. You weren't online in the sense we think of it today. You were on CompuServe or on one specific BBS – and that was it. As for e-mail, until 1993 you could only send e-mail or otherwise message the people who were users of the particular computer bulletin board service you were on.

And it gets worse. If the bulletin board was outside your local phone exchange, you'd be charged long distance for as long as your modem was connected. Before the Telecommunications Act of 1996, it cost a lot to make a long-distance call and stories were not uncommon about people first discovering the computer bulletin board world to run up multi-hundred and even multi-thousand dollar phone charges their first month online.

If all this sounds ridiculously clunky, it was. And yet the experience of being able to meet new people, take part in discussion boards, chat and send e-mail (limited as it was), access and contribute articles and graphic files, get shareware, and play (very primitive) online games was so compelling that by 1993, 3 to 5 million users had discovered the BBS world and they'd done so mostly by word of mouth. There was virtually no advertising for bulletin boards, and the news media paid no attention to them at all.

Boardwatch Magazine, as its name suggests, was the monthly chronicle of the computer bulletin board world, and ONE BBSCON was its annual industry conference. The industry was primarily made up of modem manufacturers, the makers of computer bulletin board software for people who wanted to operate their own BBS, computer bulletin board owners some of whom were trying to figure out how to make money with them, ardent hobbyists, and a handful of technology writers who occasionally wrote articles about them. One of the magazine's attractions was that it doubled as a directory of computer bulletin boards with the names and phone numbers of BBSs you could dial into.

Here's how the magazine described itself in an ad soliciting for paid subscriptions it ran in its July 1993 issue:

Connect Your PC to the World...
More than 70,000 Bulletin Boards Await Your Call

It's a new world out there. Thousands of individuals re-creating the way we live, work, play, and communicate, using a fascinating new technology to connect their personal computers in the night. In each month since 1987, *Boardwatch* has been there with the latest in who's doing it, how they did it, why anybody would want to do it, and what it might mean in the future. Find out about a new cottage industry thousands of people are leaving corporate America forever to join as independent BBS operators and online information workers in a new information economy. The people who are building tools for this new industry read *Boardwatch* every month - so should you. They know about dozens of fascinating, free, government-run information services. They learned about technology and hardware developments that drive the advances in online technology. They know the movers and shakers in the industry. Most of all they get information they need to be a part of it - and they get it from *Boardwatch*.

Then the ad went on to describe listings, local area lists, and the magazine's proprietary list of BBS list keepers.

> Short on theory, long on facts, access numbers, and contact information, *Boardwatch* delivers stories on the global Internet, commercial online services, and bulletin boards - you can dial immediately with your modem. Each issue brings you 500 to 1000 access numbers under topical lists, and geographic lists in our own selected national list of bulletin boards. We list the people who list various cities, and under various topics, all over the world, so that you can dial and download the latest list of bulletin boards from their sources. Nobody digs out more and more bullets and boards than *Boardwatch Magazine*.

If you wanted to subscribe, you had to call the office number or clip the coupon and mail your subscription order to their Lakewood, Colorado office. That's right. In 1993, the primary magazine of the online world was geared to phone and mail orders to sell its subscriptions though they did advertise their data number.

ONE BBSCON '93 expected about 2,000 attendees. Early birds who registered for the August conference before May 1st got in for $175. If you made it by August 1, $250. I registered late, so I paid the last-minute price of $325.

The conference promised 132 educational sessions and the celebrity keynote speaker was John Dvorak, a columnist for PC World who'd written such classic books as *Dvorak's Guide to PC Telecommunications* and *Dvorak's Guide to PCs*. Today these books read like Wagon Wheel Making 101, but in those days they were exactly what people needed to use their personal computer to communicate with other people. Vinton Cerf gave a well-attended session on *The Past, Present, and Future Internet*. In the 1970s, he and Robert Kahn co-wrote the technical protocols (TCP-IP) that the Internet ran, and still runs, on.

The educational sessions were divided into seventeen content area tracks. Eight of these tracks dealt exclusively with the various brands of software that ran computer bulletin boards, and four additional tracks were very tech-heavy. Only one track dealt with the Internet, with 14 of its 17 sessions addressing heavy-duty technical issues including how to connect individual BBSs to Internet e-mail and discussion boards. In addition to these tracks, there was a track on legal and social issues, a corporate/business applications track, and a government/education track, the last two focusing on special use vs. general BBSs.

This was the new world I was wading into. Me, who didn't have a modem, had no idea how to connect a modem to a computer even if I had one, and had only once in my life seen online-generated content, and that was on a computer screen in my friend's office/cabin in the backwoods of Humboldt County.

Chapter Seven

Novus Mundus

"Sometimes I think it's a bit of an advantage to not know that much."

– Paul McCartney
commenting on the fact he never learned to read music

It's very possible that when it came to the online world I was the least informed and least experienced person of the two thousand-plus people who attended ONE BBSCON '93. I didn't talk to every attendee of course, but I did talk to many people and went to an educational session in every time slot over the conference's three days. In that time I didn't encounter a single person who was at my Version 0.0 level. In this way, I was no different from over 95% of the American population when it came to the online world.

The cover of my notebook from ONE BBSCON '93.
I filled all 80 pages in three days.

That said, I also didn't encounter anyone who seemed to have even an intermediate grasp of marketing and I met no one conversant with basic direct-response advertising principles. I didn't get a hint of that from anyone speaking from the podium either, even though there was an entire track called *How to Make Money with a BBS* with sixteen individual educational sessions in it.

I went to as many of the sessions on this track as I could and nowhere did I hear anyone talk about testing, about keying ads so you knew which ones paid and which ones didn't, about cost per inquiry, cost per sale, or lifetime value of a customer. This baffled me because trying to run a subscription business, which in essence is what a BBS is, without at least attempting to know these numbers is like flying an airplane in a thick fog through a narrow mountain pass without instruments. You might make it, but only through pure dumb luck.

Up until that point, customers found computer bulletin boards by word of mouth and just showed up. If you were a computer bulletin board operator in a densely populated area, running a service competently with interesting content and entertaining things to do on it, and with enough modems so users didn't get busy signals, it was a bit like selling water in the desert. It was hard to lose. It was like what it must have been running a speakeasy during Prohibition. If you had the booze – in this case, a way for people to access the online world – and did a halfway decent job of it, people would find you.

Many BBSs were run by hobbyists with little or no ambition to make money, but if you could make your proposition attractive enough, you could sell subscriptions or memberships and some did by providing an abundance of services – chat, games, shareware, messaging, and GIF files. That is if you could take credit card payments.

One of the tracks at the conference was how to get a credit card merchant account so that you could accept credit card payments. This was not an easy thing to do in 1993 for businesses that didn't have a retail presence and didn't sell tangible products. In fact, it wasn't even that easy to get a credit card in those days. Bettina, who was well into her adulthood, owner of an apartment, and former co-owner of a restaurant, didn't have one when we met. As I pointed out in an earlier chapter, the ability of supermarkets to take credit card payments wasn't a thing until the early 1990s.

After discovering that no one at this conference knew the first thing about marketing, my next big revelation was finally grasping what the excitement about the Internet was all about.

One of the educational tracks at the conference was devoted to the subject of the Internet. This was cutting-edge stuff for these computer bulletin board operators, so much so that out of the seventeen sessions, only one addressed the topic of making money or, in the parlance of the time, *commercialization*. The rest of the sessions were purely technical and absolute Greek to me.

In 1993, BBS operators were interested in the Internet only insofar as their users were clamoring for Internet-friendly e-mail addresses because right up until this time most people online had no way to send messages to people who were not registered users of the BBS they were on. CompuServe led the way by giving its users full-fledged Internet e-mail access in 1989. This meant Compuserve customers were the first BBS users to be able to send and receive Internet e-mail, something previously only people in the military, the government, or academia could do.

Giving BBS users access to the Internet's global e-mail system only became possible after the National Science Foundation (NSF), which managed the Internet for the federal government, removed the long-standing rules prohibiting commercial activity on the Internet, a rule strictly enforced for the first twenty years of the Internet's life, from 1969 to 1989.

Compuserve had the financial and engineering might to handle this then-complex technical feat, but smaller BBSs did not and were slow to follow. By 1993, with BBS users clamoring for the wonders of Internet e-mail, it became a business necessity for BBSs, large and small, to figure out how to join the Internet e-mail party. Thus, setting up Internet e-mail gateways was the focus of the sixteen-session Internet track at the conference.

Remember in 1993, there was no Amazon and hardly anyone was selling anything online. There were no search engines. Pages did not have graphics on them. There was no streaming audio or video and the World Wide Web we know today had only been in existence for a handful of years. And no one had yet made a graphic user interface, a browser, for it.[12]

In 1993, the Web, which was the new kid on the Internet block, represented only 1% of Internet traffic and by one estimate there were only 623 websites on the Internet total. Now today, as I write this in 2024, there are, on average, 2,700 new websites created *every minute*. Of the over 1 billion websites now in existence, over 26,000,000 are online stores, aka e-commerce sites.

---

12   If you're confused about the difference between the Internet and the Web, the Internet is the basic infrastructure everything runs on and it was launched in 1969 under the name ARPANET. Think of the Internet as the train tracks and things like e-mail and web pages are like trains running on top of it. The Web, which runs on top of the Internet and which most people think of when they think of the Internet today, was actually not launched until December 1, 1990. Previous to the Web, the Internet was text-only and the experience was largely made up of sending and receiving e-mails and transferring files. The genuinely revolutionary thing about the Internet is that sending messages and files could be done globally and, for all practical purposes, instantly to and from any computer connected to the Internet network. Nothing even remotely like this had ever existed before in human history.

At ONE BBSCON '93, the one-and-only Internet commercialization education session was led by Lloyd Brodsky and one of the members of the panel was Mark Graham. Mark was one of the people who operated the Internet connection (Sovam Teleport) between the U.S. and the Soviet Union that kept a free flow of information between the two countries during the 1991 coup when all other modes of communication were shut down.

Mark also co-founded and led PeaceNet, one of the first large-scale non-military, non-academic online communities on the Internet. In 1993, *MicroTimes Magazine* named him one of the 100 most important people in personal computing, which put him on the same list as Bill Gates and Steve Jobs, for his contributions to the effort of figuring out how one could do business on the Internet. He was going to become an important person to me, but that came later.

The question the panel considered was, "Can the Internet be commercialized?" Not, "How can the Internet be commercialized?", but *Can it be commercialized at all, ever?* At the end of the session, the panelists were split. Half said they believed that the Internet would never be successfully commercialized. The other half said they believed it could, but none of them could suggest a path as to how that was going to happen. The date was August 27, 1993. That was the state of the art of thinking on Internet commercialization at the time.

"Commercialization" has a negative connotation to some people. It calls to mind websites that are jammed with a dozen or more garish ads, which I don't imagine is an appealing experience for anyone. What we meant by Internet commercialization in those days was twofold. First, how to finance building the network so that everyone could have a fast connection to it in their homes. And second, how the people who produced content for the Internet would make a living. In August of 1993, the top online industry people who were focused on these challenges could not suggest plausible ways of making these things happen.

To put this in historical context, when the Internet was first created in 1969, it was called ARPANET. It was a Department of Defense project and it had two purposes: 1) to create a resilient communications system that could survive attack, including nuclear ones and 2) to facilitate communications between universities, research centers, and others who were developing "defense-related projects", i.e. weapons.

If the experts closest to the question of how to commercialize the Internet could not envision a path, why bother with the Internet at all?

For one very compelling reason:

The BBS experience, as fun as it was to the tiny percentage of the population who were enjoying it, was clunky, hard to use, and not compelling to the average person.

In the BBS world, to "go online" meant connecting to a single computer that was not connected to any other computers. That single computer became the sum total of your online experience. You were in essence on a digital island. You could only access the content that was on the one computer you were connected to. You could only chat with people who were on that specific computer. You could only send e-mails to people who were registered with the BBS that ran on that computer.

In stark contrast, once your computer was connected to the Internet, it was connected to not one computer but a network that connected you to all the computers on the Internet. If that was ten computers or ten million or ten billion, it didn't matter. You were connected to them all regardless of where you lived or how physically far away they were from you.

This was a complete revolution in computer networking. Previous to the emergence of the Internet, computer network providers designed their systems to make their users captive to their proprietary network. If your computer was on one network, you could not communicate with computers on other networks. The Internet turned this preposterous situation on its head. With the Internet, you had access to *all* the goodies on a vast global network from a single "entry ramp," your local Internet service provider (ISP).

Once connected to the Internet, accessing the content of a computer (now we think of them as servers or websites) in Japan or Brazil or some other far-flung place around the world was functionally the same as dialing up a single local computer bulletin board.

Instead of dialing, connecting, and paying for the connection to a single server and then having to disconnect and repeat the process to get on another server, once you were on the Internet all barriers of distance, phone charges, and only being able to access one server at a time, were annihilated.

However, the Internet of 1993 had a lot of limitations.

First, access to the Internet was not easy to arrange. If you worked for the military, a large corporation, or a university, they *might* provide you with a connection. If not, you were out of luck.

Second, once you were on, even though there was a lot of content, it wasn't organized at all. You had to know in advance what you were looking for and where it was. No card catalog. No directory. No search engine. Imagine all the books ever published in the history of the world dumped into a big pile. It's all there, but good luck finding anything.

Third, the experience was text-only which might be enough for some purposes, but did not provide compelling competition for things like computer games, television, CDs, and VHS tapes (DVDs were still a few years away).

So on the one hand, we had the computer bulletin board (BBS) world, from the giant CompuServe to a hobbyist's two modem board. Anyone with a modem could play even if the experience was clunky. On the other hand, we had the Internet which eliminated key parts of the clunkiness but replaced it with significant and, at the time, seemingly insolvable problems: lack of access for regular people, total chaos once they were connected, and an experience far more primitive than the computer games and TV consumers were used to.

Something had to give, but in August of 1993, the people most engaged with the online world could not imagine what that something was.

However, two things were crystal clear to me.

First, with no advertising, no public relations, and no media hype, a small but intrepid slice of the U.S. population had managed to acquire a personal computer and a modem, and were enjoying themselves playing on computer bulletin boards despite all their clumsiness and limitations. What would happen if the experience were smoothed out so non-technical people, the vast majority of the population, could play too?

Second, everything that was in print (and remember, online was still primarily a text medium) could also be online, and once online, logistical issues and costs like printing, binding, and mailing would be gone. As someone who at the time was regularly writing four and sometimes five-figure checks with my own rent and grocery money to put a single message in the mail, my mind boggled.

I didn't understand much but I did grasp this one thing: If just 10% of my current customers got on e-mail – and as far as I knew then, 0% were – that alone would more than pay for me to take the considerable time and energy needed to master this online thing. If nothing else, I would be able to send additional messages to that 10% for free. Also intriguing, I'd be able to reach them through an additional marketing channel, always a good thing. This is sometimes referred to now as multi-channel marketing.

If my thinking sounds like it was "primitive" and my aspirations lacked grand ambition, consider the story of the founding of Apple Computers. Steve Jobs receives the credit for being the creative spark behind Apple's marketing, and rightly so. However, the original go/no-go decision to start the company in the first place came from a rough, back-of-the-envelope market analysis by Steve Wozniak.

The question that launched Apple was: "Given all the ways we can make money right now, should we invest our limited time and money in creating and selling an assemble-it-yourself computer kit for hobbyists?" Given the extent of their resources at the time, and the very real consequences of getting it wrong, the question was not trivial and the answer was not obvious.

Here's how *Woz* solved it. When he and Jobs were seeking funding, they showed the Apple II prototype to VC Don Valentine. Valentine asked them how big the market was, and *Woz* said a million units. When asked how he came up with that number, he said there were a million ham radio operators, and PCs were a bigger deal.

My back-of-envelope calculations aside ("if just 10% of my customers got on e-mail"), I didn't have the first clue about what I was going to do with all the information I had gathered at the conference. I found myself back to where I was when I was devouring stacks of books and articles on direct marketing with no place to plug the information into.

In short, I was all dressed up with no place to go, yet.

***

This chapter is dedicated to the memory of Jack Rickard founder of Boardwatch and ONE BBSCON (July 24, 1955 - August 31, 2020).

Jack was a fine man, razor-sharp, and generous with his gifts. I learned many years later that his first job out of the Navy was as a tech writer. ONE BBSCON evolved into ISPCON when BBSs either turned into Internet service providers (ISPs) or ceased to be. In 1999, he got a $40 million payday when he sold the business to Penton Media.

Jack spent the next chapter of his life back in his hometown of Cape Girardeau, Missouri collecting cars and airplanes distilling whiskey at home under a dubious "home-use" legal theory, and pursuing countless DIY projects including making his own electric cars which led to him founding EVCCON (Electric Vehicle Conversion Convention .

Chapter Eight

Somewhere Over the Rainbow

"Lord, that Hollywood train, forever coming 'round the bend!"

- James Baldwin

When I got back to San Francisco, after the ONE BBSCON conference in late August of 1993, I finally knew why some people found the Internet so intriguing. I also knew that it was not ready for prime time for two fundamental reasons: 1) too few people had access to it or even a path to getting access and 2) the content was limited, disorganized, and unsearchable.

On the other side of the equation, the BBS world, which the mainstream news media, even tech industry magazines, all but completely ignored, was vibrant and thriving. But it had limited appeal to anyone who didn't like to tinker with computers and modems. It's not a perfect analogy, but think of the difference between ham radio operators and people who just want to turn on their car radio and search for a station.

When it comes to consuming media, the overwhelming majority of people just want to push a button or two. More than that and you lose them. Based on this standard, the BBS world in 1993 was also not ready for prime time.

BBSs then were in the process of becoming gateways to Internet e-mail. Even though there was a small group of techies conversant with now-obsolete Internet functions like Gopher and WAIS (these were early attempts at search engines and menu-based systems for publishing content), for the vast majority of people, especially people writing about the topic, the Internet meant e-mail. When *U.S. News and World Report* put a Doonesbury cartoon on its cover of a long-haired hippie-esque dude surfing on a modem with his t-shirt and shorts emblazoned with "8 MB RAM", that was the Internet they were referring to.

But Internet e-mail was far from ubiquitous. Looking at the speaker list from ONE BBSCON '93, only about half included an e-mail address in their contact info and many of those were using CompuServe.com, EDU domains, or well.sf.ca.us. Speakers that had a congruent e-mail address with the name of their company were few and far between. Internet e-mail was still so rare that it was possible to write people like Bill Gates and Steve Case of AOL (if you knew their addresses), and sometimes get a personal reply. I wrote Steve Case and did.

Bound up with the issue of the Internet was the then brand-new world of digital media. In 1993, *Time Magazine* devoted three covers to digital media.

The February 8, 1993 issue of *Time* had the cover headline "Cyberpunk". It was no doubt inspired by *Wired* which had launched the previous month. The *Time* article portrayed a coming world of "virtual sex, smart drugs, and synthetic rock and roll." Like *Wired,* the article sensationalized digital media, generating much heat with near-zero light.

The April 12, 1993 issue of *Time*, "The Info Highway", featured a world that never came to be, but disappeared billions of development dollars, namely "interactive television". It was a technology intended to connect consumers to corporate-produced and controlled content via their television sets. Warner Cable, Time Warner, and even American Express experimented with services in Orlando, Florida, and Columbus, Ohio.

The September 27, 1993 issue, "Attack of the Video Games", was the most insightful and useful of the three. It pointed out the still underappreciated fact that computer games had eclipsed Hollywood in revenue, audience size, and involvement.

A year later, maintaining its "digital media is weird" stance, *Time* ran only one cover story on the subject, "The Strange New World" of the Internet which appeared in its July 25, 1994 issue. It referred to "battles on the frontiers of cyberspace" and asked the question, "Is there room for everyone?" Again, much heat and no light.

In 1995, *Time* devoted only two covers to digital media. The June 5, 1995 issue was an embarrassing hosanna to Bill Gates "Master of the Universe", the title of the article, in which he was praised for "having conquered the world's computers." The August 21, 1995 "Cyber War" issue breathlessly announced, "the U.S. rushes to turn computers into tomorrow's weapons of destruction."

In short, anyone looking to the news media for useful insight or guidance about the Internet and the digital world was likely to not only miss the plot entirely but also be actively misguided. Books on the subject were few and far between, with 99% of them being technical guides, and the majority of them rushed to print with little care for accuracy.

One exceptional book from this time deserves mention, Brendan Kehoe's (1970-2011) *Zen and the Art of the Internet: A Beginner's Guide.*[13] Reading it today provides a good picture of how computer people regarded the Internet as of January 1992. The preface tells it all and also contains a genuinely prophetic statement:

> *"The composition of this booklet was originally started because the Computer Science department at Widener University was in desperate need of*

---

13   For the original text of *Zen and the Art of the Internet: A Beginner's Guide,* see: https://legacy.cs.indiana.edu/docproject/zen/zen-1.0_toc.html

documentation describing the capabilities of this "great new Internet link" we obtained...

One warning is perhaps in order--this territory we are entering can become a fantastic time-sink. Hours can slip by, people can come and go, and you'll be locked into Cyberspace. Remember to do your work!"

Because Kehoe made the text of his book available for free, it was one of the most widely shared and read introductions to the Internet for years. There's no telling how many people Kehoe's book successfully took by the hand and guided onto the Internet. Readership could have been as high as one million or more and is known to have included a young Sergey Brin, one of the co-founders of a once-great search engine called Google.[14]

In summary, the Internet had vast potential, but was nowhere near ready for prime time. The existence of BBSs was tangible proof that people liked to spend time and do things online but they too were nowhere near ready for prime time. A vision of how these intractable barriers were going to be resolved, despite all the cover stories, articles, papers, and conferences, was nowhere to be seen.

By good fortune, I happened to be living in the absolute best place on earth to take a stab at sorting these things out. At the time, San Francisco was the Detroit of digital media. It was where the "cars" were made. Silicon Valley and Stanford University were just a short drive to the south. The University of California-Berkeley, a hotbed of Internet software infrastructure development and the launching pad for the careers of thousands of computer-savvy engineers, was just across the Bay. San Francisco, which always

---

14   Like Prometheus, the Greek god who stole fire from the gods on Mount Olympus to give to humanity and was punished by Zeus, Kehoe was struck down on the last day of 1993 seriously injured in a car accident sustaining a brain injury that robbed him of his speech. He recovered only to die from acute myeloid leukemia and its treatment eighteen years later, in 2011.

attracted artists and visionaries, was the place where the world of digital media was assembled from "parts" made from all over the world.

Some of the "parts" included the GIF file developed by CompuServe (Columbus, Ohio), the JPEG developed by international collaboration at the instigation of the International Organization of Standardization (Geneva, Switzerland), the MP3 developed by the Fraunhofer Institute for Integrated Circuits (Erlangen, Germany), and the PDF by Adobe Systems (San Jose, California).

The effort to figure out how to digitize media – graphics, audio, and video – and make it available to all via a global computer network eventually became a worldwide effort, but at the time, San Francisco was *the* place where it was all coming together. However, in 1993, San Francisco's digital media industry for all practical purposes paid no attention to the Internet. Their focus was on creating digital interactive multimedia, but at that time the vision was that it would be delivered *exclusively* by CD-ROM, a medium much-hyped by Bill Gates and Microsoft.

As late as 1994, Gates was so devoted to his vision of the CD-ROM being the distribution channel for digital media that he posed for a promotional photo suspended from a crane in a forest showing that one CD-ROM could contain as much data as two piles of print outputs, each pile being the height of a tree.

As I quickly came to learn, people working in the fledgling digital media industry had next to zero interest in the Internet as a potential distribution channel. Their outlook was: "We can barely transmit photos and we can't deliver audio or video with current modem speeds. Why should we waste time developing content for the Internet or even thinking about it?"

They were correct in one sense, but very wrong in another. At the time, it was true that access speeds for consumers provided an experience as exciting as watching paint dry. I remember the first time I showed Bettina a photograph in the process of downloading from the Internet on my personal computer's screen. The process was so painfully slow, she lost interest halfway through and could not fathom what I was so excited about.

I knew then, and still know, nothing about computers other than how to use them to write and surf the Web, but I did know one thing: Everything – graphics, audio, and video included – could be reduced to 0s and 1s. A charming scientist and amateur clarinet player, Claude Shannon had sorted this out in 1948 for Bell Labs when they were trying to figure out the best way to accurately transmit messages without them falling apart in transit over long distances.[15]

I knew nothing of this history, but the 0s and 1s thing was in the air and every reasonably well-informed person knew about it. I kept asking multimedia developers, which is what they were called then, if it was true that all media could be reduced to zeros and ones. They said yes, it was. Then I asked them: "If we had faster pipes, then we could transmit audio and video with no problem, right?" They agreed, but the question was as practical to them as if I'd asked: "If we had wings, could we fly?" The leap in modem speed necessary for consumers to transmit video was inconceivable to them.[16]

No one in San Francisco's 1993 cutting-edge digital media industry who I spoke to, and I spoke with a lot of

---

15   The new and improved modem of the 1993 season provided 19.2 kilobits per second. As of this writing, the average download speed for a user in the U.S. is 219 megabits per second. A megabit per second is more than 1,000 times faster than a kilobit per second. So in making the leap from 19.2 kilobits per second to 219 megabits per second data, transmission speeds today versus then are roughly 20,000 times faster.

16   "0s and 1s" are the binary code building blocks that allow computers to store and process any type of media, like images, videos, audio, and text.

people, saw any relevance of the Internet to their work. Most of them didn't have an e-mail address on their business cards. This state of affairs would persist well into 1995. The view was so entrenched that in late 1994 when I commissioned the head of the Bay Area's Digital Video Special Interest Group (SIG), Hank Duderstadt, to write an article on the potential of video on the Internet for the first issue of the *Internet Gazette* (a tabloid I published in 1994 and 1995), he spent more time explaining the technical barriers than imagining the possibilities.

His position made a certain amount of sense. Modem speeds and computers themselves were much, much slower then. In effect, I was asking people who were in the trenches hand-assembling Model A cars (Henry Ford's first product) to imagine tooling down Interstate 95 from New York to Miami at 60 miles per hour.

In the Model A days, cars were so unreliable that each one came with an elaborate toolkit including an adjustable wrench, grease gun, oil can, and pliers because constant breakdowns were expected. Programming multimedia content for personal computers was not much different.

Still, I could not help believing that video on the Internet was not only technically possible, it was also inevitable. I was just a little early. About twelve years early. I received support for my faith from an unexpected source, a history published in 1992, called *The Story of Telecommunications* by George Oslin. When the book came out, Oslin was 92 years old. He began his working career as a reporter and ultimately became director of public relations for Western Union at a time when the sending and receiving of telegrams was part of everyday life.

Among many things, he invented the singing telegram, an idea which first earned him the scorn of fellow executives, but became a source of revenue and endless free publicity for decades for the company.[17]

In addition to being a reporter and master of publicity (a benign version of Edward Bernays), Oslin was an avid historian. Using his position at Western Union, he assembled an enormous collection of documents and first-hand accounts, not only of the history of the telegraph but of all message delivery systems. He interviewed the last living Pony Express rider. The riders were men who transported the mail via horseback between far-flung western communities. He also interviewed Thomas Edison who started his career as a telegraph operator and, among other things, accidentally invented audio recording while attempting to create a method for relaying telegraph messages.

The book's stories fascinated me and three things especially stood out:

First, before the telegraph, there had been no change in telecommunication speed from Sumeria to the age of Queen Victoria. It never got better than relayed semaphores (flag waving) and men on fast horses.

Second, the technical challenges of physically wiring the world, first for the telegraph and then for the telephone system, were nothing less than epic. For example, imagine the challenges of laying the first transatlantic undersea telegraph cable.

Third, when telecommunications changes, the world changes. The telegraph, radio, and television all radically changed the details of human life.

I had the great pleasure of corresponding with Mr. Oslin and he was very encouraging of my work on the

---

17   You can still order singing telegrams today, but not through Western Union. They suspended the service in 1974, but you can find singers on the Internet who will physically show up at the physical address of the person of your choosing and deliver a sung message.

Internet side of things which then was little more than thinking, imagining, and talking to people.

My conclusion was that if people could figure out how to lay cable from one end of the Atlantic to the other and send messages across it and then go on to build an international telephone system, surely someone, or group of someones, would be able to deal with the technical challenges of getting fast digital pipes into everyone's homes.

Reading Oslin's panoramic history of telecom gave me the sense that the technical improvements and expansion of the network were simply not the barriers everyone expected them to be.

The reason I was able to put my questions and ideas in front of so many multimedia people was because, by craft and luck, I'd made an important connection in that world very soon after I arrived in San Francisco in 1990. *The New Fillmore*, a wonderful print newspaper that documented events in my neighborhood, wrote a profile of a local resident named Hal Josephson who was an important pioneer of the multimedia industry.

Hal, with underwriting from Microsoft, Apple, IBM, and New Media Magazine was a kind of Johnny Appleseed for multimedia on the personal computer. He organized conferences around the country to showcase the ideas and work of pioneering multimedia producers to ad agencies, public relations firms, and in-house corporate media departments. The pitch was that if you were producing print, audio, and video for yourself and clients, you should start learning how to produce interactive digital multimedia too because that's where everything was headed.

After I read the profile on Hal, I looked him up, introduced myself. I let him know that if he ever needed any copywriting he could call me. Coincidentally, he needed some right away. This bit of copy I wrote for a brochure will give readers a good sense of the state of digital multimedia in 1992 (the occasional odd language and capitalizations, no doubt created by me, are reproduced as it appeared):

Who Should Attend?

Interactive '92 is designed for professionals whose business or career is linked to media and communications. This includes:

* Ad agency personnel
* PR representatives
* Sales and marketing executives
* Directors of Corporate Communications
* Communications Managers
* Corporate Trainers

Information about Interactive Multimedia that You Can Use

If you've been sitting on the multimedia sidelines or experimenting and wondering how to take the next step, this workshop is just what you've been waiting for. Here's an opportunity to meet seasoned multimedia producers, and find out firsthand. What's really going on. No pie-in-the-sky projections, no rhapsody about the future of media. Just real-life applications that have helped corporations, ad agencies, schools, and small businesses educate, inform, sell, persuade, and entertain.

Real Life Case Studies

* You'll meet knowledgeable multimedia producers
* You'll see their projects in action
* You'll hear their unbiased opinions about hardware & software
* You'll learn how they address the practical issues of budgeting

My interest in the idea of interactive media went back to 1988 when I was working with Bill Markle in the Film Center Building. I saw an ad for a conference at the Javits Center on the subject of creating and marketing VHS tapes to the home market. VHS players had finally become common items in the home with Hollywood providing the first wave of content by converting its feature films to tape. With lowered costs of production thanks to the transition from film to video, some speculated that there was a business in making and marketing "how to" videos on a wide variety of subjects.

Since reaching the market was the big challenge (then as now, making the videos was the easy part), the advice to producers was to partner with special interest membership organizations, for example, the Audubon Society to sell VHS content on bird-watching or with magazines, like Field and Stream to sell videos on fly fishing and hunting. The idea of lining up distribution before production made good sense, but the business itself didn't thrill me.

However, there was one speaker at the Javits Center conference who talked about the prospects of interactive TV which I did find interesting. She was from MIT and in retrospect, she was probably from the then-new MIT Media Lab. I'd never heard anyone put the words "interactive" and "television" together before. As someone who had recently discovered the wonders of video editing through working with Bill Markle, I was intrigued. I found the ability to pause, rewind, fast forward, and then cut and move clips on video tape to be an intoxicating activity.

Of course, anyone with one VCR could do the first set of things and anyone with two VCRs could rearrange existing video, but having a video controller built for the task made it easier, more enjoyable, and the final product much better. The notion of interactive television gave me the idea of how wonderful it would be to be able to stop a broadcasted program in midstream and replay part of it, skip over the boring parts, and then easily save and organize the specific bits that I liked.

After hearing this one talk, a few years later, when I got to San Francisco and discovered that there was a bona fide interactive media expert in my neighborhood, Hal Josephson, I jumped at the chance to meet him.

As I soon learned, deep-pocket clients could afford to commission custom interactive digital creations for booths at trade shows and other high-stakes sales-oriented purposes and these one-off productions were the meat and potatoes of the multimedia production world at the time. However, the dream of many in the industry was to create personal computer-based interactive entertainment for the masses and sell thousands if not millions of units of a single title.

The fantasy, and I call it that because that is what it proved to be, was stoked by people at MIT's Media Lab, its friends at *Wired*, and countless poorly informed reporters. The fantasy went like this: "Occupying the space between computer games and Hollywood blockbusters, a new genre and industry is going to emerge: interactive movies on CD-ROMs. And this industry and the people in it are going to make billions."

This excerpt from an unsigned article on RedSharkNews.com provides a good example of the state of the art of CD-ROM interactive entertainment at the time:

> One of the much talked about CD-ROMs of this era was the 1964 Beatles movie A Hard Day's Night, released by Voyager in 1993. The QuickTime movie ran on a postage stamp-sized screen at 12 frames per second, but this was partly compensated by the fact the screenplay ran alongside it. The disc also boasted 'interactive menus for finding favorite songs and scenes, and special extras

like an interview with the director and the theatrical trailer'. It retailed at the time for $40 – equivalent to around $70 today. A lot to pay for a 30-year-old black and white movie running at half speed in a minuscule frame.

In a world where people could go to the movies for less than $5 or rent one for the entire family for even less than that, a $40 price point (really more like $100 today) made little sense. It was true that popular computer games were more expensive than multimedia CD-ROM titles, but a computer game user might spend 50 to 100 hours or more on a single game versus a one or two-time viewing of a multimedia CD-ROM.

In short, in 1993 I had found the people who were on the cutting edge of San Francisco's digital interactive multimedia industry, but in my estimation, the business model they were envisioning was ultimately leading them off a financial cliff.

Chapter Nine
Digital Diplomacy

"We have it in our power to begin the world again.
A situation, similar to the present, hath not happened since
the days of Noah until now."
– Thomas Paine

When I was a tech writer for Bankers Trust and later for First Boston, besides doing my own work, I was also the go-between for the traders and the computer system design people, two radically different cultures.

Traders are hyper-vigilant and move fast. It's the nature of their business. Arbitrage opportunities, the chance to buy low and sell high before other people catch on, sometimes only last for hours, sometimes just for minutes.

In contrast, computer folks, like engineers who design bridges, are measured and methodical. They have to be. If they get things wrong, very bad things can happen. An incorrectly designed bridge can fall down. An incorrectly designed trading system that adds or eliminates a zero or two from a transaction can have financially catastrophic results.

In terms of the pecking order, traders and the trading department were kings of the hill. As a lowly tech writer, I had to convey computer information to them quickly, simply, and clearly. One take. And then get lost. They barely tolerated the computer folks.

Now, a few years later, I found myself in a similar, but different situation in San Francisco.

Online people, which included the Internet folks, were essentially engineers. They were theoretically interested in making money, but when you looked at the content of their conferences like ONE BBSCON and other industry get-togethers, it was 95% tech talk, and understandably so. The systems had to work.

The multimedia folks worked with engineers but they were primarily artists who were focused on the *content* they were creating. They also had serious dreams of their work eventually evolving into an interactive digital Hollywood with all the bells and whistles that would come with it.

I liked the Internet people and I liked the multimedia people too. They were both infused with the spirit of discovery and adventure and with very few exceptions were always willing to take the time to talk with me and answer my questions. They talked with me, but they almost never talked with each other.

I found this very strange.

Back from ONE BBSCON '93, I became a digital diplomat and worked to bring the two groups together. I had modest aspirations. I reckoned that if the Internet took off, and just 10% of my current customers became online users, it would make it possible for me to "tell and sell" more. I was, and still am, simple-minded on this point.

As a business owner, I want customers, people who buy things from me. The closer I can get to my customers through frequent communication, and the better I am at creating new customers, the better things will be for me and my business.

The "nuts and bolts" of business is generating leads, following up on leads, closing sales, developing new products for existing customers so you can sell them more, and retaining the customers you have.

Any person, business, or non-profit that is not focused on these things on a daily basis is missing the boat entirely. I know this sounds ridiculously basic, but I doubt that 1 out of 10 (probably less than 1 out of 100) "in business" actually thinks this way and lives it.

Pre-Internet, lead generation meant running ads in newspapers, in magazines, on the radio, on TV, and using direct mail. "Free" publicity meant being featured in one of these media or arranging to speak and/or display your wares at a relevant gathering.

Closing sales meant getting face-to-face using the telephone, or selling to a live audience in a room. Sales could also be done by print catalog or physical mail. You could improve your communications with prospects and customers by including audio cassettes or even video cassettes in your mix. There were a lot of options and smart businesses (the 1%) used as many of them as they could manage.

People in the direct response world are exponentially more aware of these tools than other types of businesses. It's in their DNA to try as many media as possible and because they track their efforts and do arithmetic, they learn which media work for their offers and which don't. They also learn which messages work and which don't. They are agnostic. They have no "favored" media, method, or even message. If it works, they use it.

When I tried to explain this mindset to my new Internet and multimedia friends, most of them either didn't get it or were openly aghast. "You mean modeling our work after junk mail and infomercials?" This did not appeal to people who had aspirations for changing the world or getting an Oscar for best interactive multimedia production.

Here are the messages I tried to convey to them:

1) If the Internet is ever going to be competitive with existing media like radio and TV, it has to have a robust "multimedia" component. Internet folks are going to have to figure out how to make audio and video work online. Period.

2) The Internet, like it or not, is the ultimate direct response media. Why? Because you can track the things every marketer needs to track: How many people see your ad, how many people interact with it in some way, and how many people buy.

3) The Internet is the dream distribution channel for multimedia developers, And they should learn as much as they can from the Internet people and collaborate with them at every opportunity.

In 1993 neither side was convinced.

If the Internet and multimedia people didn't get what I was saying, I knew or thought I knew, one group that would. I assumed that would be folks in the direct response industry. After all, they're the people who produce the direct mail pieces, the radio and TV spots, and direct response newspaper and magazine ads for clients and their own projects. Surely, they would be enamored of a media that had simple tracking baked into the cake.

I soon discovered that in 1993, the overwhelming majority of people working in the direct response field also didn't get it and had no interest in getting it. They had mailings to prepare, lists to broker, spots to produce, and media to buy. The very much not-ready-for-prime-time Internet held zero interest for them.

But I knew one person who might be interested: Dan Kennedy. He was one of the key people who introduced me to the nuts and bolts of direct response, specifically how to use it and use it profitably especially if you were serious about succeeding and not just spending somebody else's money. Dan was all about reality and had zero time for theory or posturing.

Probably the simplest way to grasp Dan is to go to Amazon and look at the books he's written, or rather the long shelf of books he's written. Then to form an accurate impression of the scope of his work, you have to envision that this mass of writing is the small tip of the very large iceberg of what he has learned, and continues to learn, from real-world practice and experience.

When Dan was younger, his older colleagues nicknamed him "The Professor" because then, and it's 100x the case now, he'd made the effort to learn more about the practice of direct response advertising and how to run direct response-fueled businesses than any living person, and quite possibly anyone who's ever lived.

Anyway, as soon as I got back from ONE BBSCON '93, in addition to keeping my core business afloat and running around being an unpaid evangelist to Internet folks telling them, "You need to think about multimedia," and to multimedia folks, "You need to think about the Internet," I started a newsletter. It was delivered by fax (in 1993, the "cool kid" on the block), and it contained my musings about the evolution of digital online media and its potential repercussions for direct response. I had one subscriber: Dan Kennedy.

A few months later, I received a message from Subscriber 001: "I'd like you to come to Phoenix and talk to my top clients about this online thing." To me at the time, it was the equivalent of Michael Jordan inviting me over to his house to play some hoops.

It was on.

Chapter Ten
Out of the Desert

"I will prepare and someday my chance will come."
                                    – Abraham Lincoln

It's November 1993 and I'm getting ready for my talk in Phoenix to Dan Kennedy's top clients.

It's one thing to casually chit-chat with people about futuristic scenarios. It's another thing to be a speaker on a then-arcane subject to an audience of very focused, very driven, result-oriented business owners who know and care nothing about what you're talking about and who had paid $4,995 to attend the meeting. (Spending time in a Dan Kennedy audience was not then and is not now a cheap date.)

Fortunately, I was not the only speaker, but still, there's a world of difference between cocktail party-level banter and standing and delivering in front of a demanding audience for an hour. Suddenly things became very real. What on earth was I going to tell these people that was going to be immediately useful or even make any sense to them?

Maybe I was delusional. Maybe I'd been obsessively sinking hours every day into attending meetings and conferences, meeting with people, researching, and reading articles about something that was never going to happen.

To the world outside San Francisco in 1993, the Internet and multimedia were a nothing burger. In fact, up until Netscape's IPO in 1995, they were a nothing burger to most San Franciscans. Even as late as mid-1995, when I attended a software industry conference and attempted conversations about the Internet with C-level executives from some of the software startups that were exhibiting, I might as well have been talking to Americans about what team was likely to win the world badminton championship that year. The idea of producing software for the Internet or even paying any commercial attention to it did not compute.

Luckily, as I was preparing for the Kennedy conference in 1993, Cecil Hoge came to my rescue again. He was the author of *Mail Order Know-How*, the book that clued me into the fact that direct response advertising was a 100+-year-old thing, not something I invented to sell seats in my speed reading classes. He had written another book, *The First Hundred Years are the Toughest: What We Can Learn from the Century of Competition Between Sears and Wards*, I'd come across while exploring the stacks of San Francisco's old main public library. It told a story from the 19th century that predicted what was going to happen next online.

As of this writing, there are only 12 Sears stores left, down from 3,500. The original Montgomery Ward went out of business at the end of 2000. At the end of the 1930s, it was the largest retailer in the United States. Combined and at their peak, these two companies were bigger than today's Amazon, because, in addition to their physical stores, which were massive, they also had mail-order catalogs in virtually every home in America. In fact, that's how they started – Montgomery Ward in 1872 and Sears in 1893. The story of these two companies, especially their early years, was 100% applicable to understanding the Internet in 1993 and its longer-term potential.

In 1880, midway between when Sears and Montgomery Ward were founded, 72% of the U.S. population lived in rural areas, and over half of the labor force was engaged in agriculture. In contrast today 14% live in rural areas and less than 1% of the U.S. population are agricultural workers.

In the late 1880s when Sears and Wards were getting rolling there was plenty of money in rural agricultural communities as anyone taking an inventory of grand Victorian houses in small-town America can see with their own eyes. The market was not only huge, it had money. What these people didn't have was access to the same goods as their urban counterparts.

Aaron Montgomery Ward and, a little less than two decades later, Richard Sears came up with the idea of selling to these folks via a print catalog. It all sounds pretty obvious now, but it was a society-transforming revolution back then. And it was entirely dependent on the convergence of several technologies that had only just come into existence, namely mass production, a national railroad system, universal postal delivery, and cheap printing. Factories, railroads, and postal systems had existed before, but on a much smaller, less distributed scale. In the second half of the 19th century, particularly after the Civil War, they exploded.

For a mail-order catalog business like Ward's and Sears' to exist, you had to be able to print the catalogs at a price so low you could afford to give them away for free. There had to be a means to deliver the catalogs to everyone who wanted one and then receive their orders and ship their goods straight to their homes. You needed the goods to sell in the first place, which is where the then-new mass production came in. Finally, you needed a population of people with spare cash to spend. Without all of these technologies and systems coming into being, there would have been no mass catalog business.

As I looked at the technology and infrastructure of the early 1990s, I noticed a similar convergence taking place. Personal computers were becoming ubiquitous not only at work but also in the home, and the machines were getting faster and better all the time. There was a fledgling online world, the Internet for academics and government people, and the BBS world (CompuServe, Prodigy, AOL, and thousands of small boards) for everyone else who was interested. Modem speeds were getting faster and faster. It was a clunky system, but it showed that people liked doing things online and naturally gravitated to it.

What would happen if all the clunkiness were taken out of the system and everyone had their own well-oiled computer attached to it? Many things we were then distributing via print, mail, and broadcast could be sent for an infinitesimal fraction of the cost and virtually instantly. The cost of communication would collapse and all kinds of enterprises, previously unimaginable because they were economically impossible, would come into being.[18]

I told this story to the attendees at Dan's event. My argument to them was that everything was lining up for a massive expansion in how businesses would be able to reach prospects and customers, one that would make many currently impossible things possible. I recommended that at a minimum they get an account with one of the large computer bulletin board services (I recommended AOL), that they get an e-mail address, and add it to their stationery and see who, if any, of their customers and vendors were already online.[19]

I added that the BBS world was intriguing, but was hampered by the fact you had to market separately to each digital "island" – CompuServe, Prodigy, etc. – with each being its own self-contained digital world. I described the Internet as the wild card, as "vast as the ocean", and that "if, somehow, it could be tamed, it would change the world."

18   A talk on this topic which I gave in San Francisco on November 5, 1994, lays all this out in much greater detail. The full text appears in *Appendix II* of this book.
19   A month earlier, I'd given the same advice in an article I wrote for the *DM News*, the magazine that covered the direct marketing industry. It was was the first article to appear in a mainstream marketing journal that encouraged marketers to get e-mail addresses and to take e-mail seriously. I called it, "Why We Will Not Have Interactive Television", the full text appears in *Appendix I*.

In November of 1993 when I gave this one-hour talk, which I conservatively spent 100 hours preparing for including all the rumination and research that went into it, I wasn't sure how that was going to happen – and neither was anyone else.

- Part Three -
Getting the Band Together

Chapter Eleven
Off the Shelf

"It does not take a majority to prevail…but rather a tireless minority, keen on setting brushfires of freedom in the minds of men."

— Samuel Adams

My trip to Phoenix accomplished two things:

First, the process of studying the history of the old mail order giants like Sears thoroughly convinced me that somehow faster computers, faster modems, and improved online experiences were going to converge and make the online world a part of everyday life. It was inevitable. I just didn't know how or when it was going to happen. My hunch was we were at least ten years away, but that it would be worth sticking with it regardless of how long it took.

My bread-and-butter business of organizing and marketing conferences for people in the mortgage industry was motoring along so I could wait it out regardless of how long it took the medium to mature.

Second, the talk I gave in Phoenix was good enough that for the next eleven years, until he sold his company, I was the only person Dan Kennedy invited to speak to his clients about online marketing. That covered the period from 1993 to 2004. Over that time, I was put in front of a few thousand serious prospects for my consulting and services, and in 2002 I used the following I'd built from that to start a conference and seminar business on Internet marketing, one that generated millions of dollars in revenue.

I launched that business, the System Seminar, in the smoking crater of the first dotcom crash when companies like AdTech were on the verge of bankruptcy. It became easy to rent apartments in San Francisco again, and the ever-unreliable news media was suggesting the Internet might be a passing fad that might not ever financially recover.

But on January 1, 1994, on the cusp of the boom that led to the crash, the idea of a commercial Internet hadn't made much visible progress since it was declared open for business back in 1989. True there was a growing amount of media hype and a lot of activity of dubious value, but the state of the Internet as a place to sell things and publish commercially was as primitive as rubbing sticks together to try to make a fire.

To give you an idea of how much business was *not* being done on the Internet in 1994, Dave Taylor, who started on the Internet in 1980, published a directory of all the companies he could find that were attempting to do business on the Internet as of November 1994. The printout came to about 400 listings. Roughly half of the companies were computer-related and less than half had a website. The rest listed only an e-mail address with a few showing arcane and now long-forgotten technologies like Gopher, WAIS, and anonymous FTP as the way for prospects to be in touch with them. By way of contrast, 30 years later there are many millions of websites that take money in exchange for goods and services.

I specifically mention the years 1989 to 1994 because, until January 1, 1989, it was forbidden to use the Internet for commercial purposes. Internet use was restricted to government (.gov), educational (.edu), and military (.mil) organizations. The first .com domain name didn't appear until March 15, 1985, when a Cambridge, Massachusetts military contractor called Symbolics, Inc. registered symbolics.com.

From 1969, when ARPANET, the forerunner of the Internet, started to 1989, twenty years later, the Internet was a business-free zone. You could not advertise on it. You could not sell products on it. You couldn't even sell access to it to consumers. Removing those restrictions in 1989 did not set off a gold rush. It was crickets. And it was crickets in 1990, crickets in 1991, crickets in 1992, crickets in 1993, and circumstances gave all the appearances of it being crickets in 1994 too. But appearances can be deceiving.

Throughout 1994, I kept reading, researching, talking to people, and attending industry shindigs in the online, multimedia, marketing, and the then-Internet-less software industries. I was in search of someone to show me the way. By early 1994, it was dawning on me that there was no one showing the way. The impression I got at ONE BBSCON '93 in Colorado proved to be accurate: The market, such as it was, was tech people talking to tech people about something they were trying to build that they didn't have a clear blueprint for. So I continued to watch, wait, and circulate.

As an example of how widely I cast my net, when I saw a notice that SIMBA Information[20] was holding a conference on Yellow Page advertising, I decided to go. I imagine that at least half the people reading this are not going to have the faintest idea what I'm talking about. Before the Internet, if you wanted to look up a local business or find a local supplier or service provider for something you needed, the *only* resource was a print directory called the Yellow Pages, distributed for free by your local phone company. If you needed to find providers outside your area, you'd have to go to the main branch of a big city library and they *might* have Yellow Pages books for some major cities.

Yellow Pages was a stodgy old business. Businesses paid for listings and it was a big money maker for the phone

---

20   SIMBA specialized in producing research and creating conferences for the media, publishing, and advertising industries on a wide variety of topics.

companies. The more you paid, the bigger the listing you got and you could buy multiple pages in multiple categories if your budget could afford it. There was nothing cutting edge or digital about it, but I reasoned that if ever there were a product that would work better online, it was the Yellow Pages. Imagine being able to look up the Yellow Pages of the entire world without having to physically go to a big library. The phone companies would save millions on printing and distributing the books and the listings would, theoretically at least, be more up to date.

At the Yellow Pages conference, only two speakers talked about the future. One had worked at or with QVC, the TV home shopping network that was only eight years old at the time. Home shopping networks, an industry that got started in 1982, were a fusion of entertainment and commerce. Consumers bought directly from the media owner, similar to what the Internet has evolved into.

The speaker also had some experience with the business side of the online world. It was either with Compuserve or an ill-fated company called CUC (Comp-U-Card) International. I wish I could remember his name because he said something profound that deserves a citation. He summed up why the online world was such a draw:

"People go online for four basic reasons: As a cure for boredom; loneliness; as a way to advance their careers; or, if they're entrepreneurial, to make money."

Another speaker caught my attention. He came from the ad agency business and was a media buyer for Hal Riney & Partners, a hip advertising agency in San Francisco that had a national reputation. Media buyers figure out where to run the ads that the people in the creative department make. They know about all the available market channels, figure out which ones make sense for the offer, negotiate prices (all rate cards are fiction), make the media purchase, and then monitor the campaign to make sure the ads actually appear and run where and when they are supposed to and then report back to the client. In 1994, their work covered newspapers, magazines, radio, TV, and billboards.

In the old days, media buying, especially for TV, was pretty easy. Make buys on the three networks – ABC, CBS, and NBC – and you were done for the day. Cable TV which made its nationwide appearance in the 1980s changed all that. Suddenly viewers had upwards of 50 channels to choose from. The speaker pointed out that we weren't that far from a world with 500 channels on our TV sets, and asked the assembled audience how media buyers would deal with a future like that, which he asserted was coming on fast.

When he finished his talk, I walked up to the speaker – his name was Rick Boyce – and introduced myself. I asked him if he'd given any thought to how to address the problem of having 5 *million* channels to choose from. I picked that number for dramatic effect not knowing that we were heading to a world with over 5 *billion* electronic channels (websites).

"Have you ever heard of the World Wide Web?" I asked. He had, but he admitted that he didn't know much about it. Neither did I, but I'd seen a demo of Mosaic and it smelled like the future. Based on my limited understanding of the technology, it appeared to me there was no limit to how many websites there could be and each website would be like a separate media channel.

A few weeks later Rick invited me to visit his office and I told him more about what I'd learned and where I thought things might be headed. He listened to my ramblings thoughtfully, the first mainstream ad agency or marketing person who did. Wheels were obviously turning. "I don't know what ads on websites would look like," I said, "but it seems to me there has to be a way to sell ads on these things."

On a high shelf in his office, Rick had a complete collection of all the Standard Rate and Data Service directories. The SRDS. There was one volume for each medium: radio, TV, newspapers, magazines, mailing lists, and miscellaneous. These big books listed all the advertising for sale along with their demographic and rate information, and were the bible for media buyers.

Rick pulled down the volume entitled "Miscellaneous". "I imagine this is where Internet ads would be listed." Neither of us realized it a the time, but this was a momentous moment for the future of the Internet. Yes, there was a place for Internet advertising in the media buyer toolkit. A serious media buyer just said so.

Chapter Twelve
Great Idea, but Who's Going to Pay the Bill?

"Nothing happens until somebody sells something."
- Red Motley (1900-1984)

As early 1994 rolled along, I felt like I was making some progress.

I'd finally met someone I was able to have a real-world business conversation with about the Internet's prospects as an advertising medium. That would be Rick Boyce, the media buyer for Hal Riney & Partners. If there was anyone else in the entire global advertising industry at that time who was willing to devote detailed thought to the question of advertising on the Internet, they'd kept themselves well hidden straight up to and through most of 1994.

No one in the advertising industry was writing about the Internet or speaking about it at conferences in 1994. At the time that made perfect sense. Despite their arty pretensions, ad agencies are advertising *agencies*. They don't get paid for making clever ads. They get paid for making ads that please their clients and their compensation is a percentage of the media buy (typically 15%). Thus a $10 million ad spend yields a $1.5 million commission for the agency. Given that in 1994 there were exactly zero Internet ad spaces available for sale, there was no immediately obvious reason for the ad agency industry to give the idea any attention. And they didn't.

Ad agency execs weren't the only people not talking about business on the Internet. Practical discussions about the subject among people employed in the Internet world were almost as rare. The Internet, despite what a few incoherent bursts of hype from the news media claimed, was still a very small network. The engineers who kept it running were all salaried people working for governments, universities, the military, or the rare tech businesses like Sun Microsystems that made money selling the hardware the Internet ran on.

To compound the program, Internet people didn't talk to San Francisco's "digerati", the people who developed multimedia CD-ROMs. In turn, the digerati didn't talk to the Internet people. Software industry people had zero interest in the Internet as a business opportunity and the leaders of the industry from Bill Gates of Microsoft, Steve Jobs of Apple, and Larry Ellison of Oracle on down publicly disparaged its commercial prospects when the subject was raised in interviews. Wall Street was even less interested than the software industry.

One shining exception to all this inactivity was Mark Graham. I mentioned Mark earlier as the person who ran the Internet track at ONE BBSCON '93.

Mark was a computer engineer who, thanks to a stint with the Air Force at their doomsday computer center at the Pentagon,[21] arrived in the Bay Area with lots of highly practical Internet experience under his belt. By 1994, he'd helped put the peace movement on the Internet with PeaceNet, ran the tech for one of the first Internet service providers that served the public (IGC.org), and managed technology for Sovam Teleport, the Internet link that kept the U.S. in communication with the former Soviet Union in the middle of its brief but chaotic 1991 coup. As a testament

---

21   You may be comforted, or appalled, to know that one of the scenarios the Pentagon is prepared for is an all-out total nuclear war and they have plans and gear to keep their communications systems running despite one.

to his Internet chops, when, in the pre-Web Internet era, AOL wanted to create a gateway to the Internet for its users, he was the guy who made it happen. And Mark lived in San Francisco just blocks from me.

I mentioned earlier that one of the central figures in San Francisco's multimedia industry, Hal Josephson, had his office near me. It was one and a half blocks away to be precise. Up Fillmore Street one block and then half a block down Sacramento Street and I was there. Mark lived a bit further away. I had to walk a whole five blocks to get to his apartment on Broderick.

After some back and forth with Mark on the telephone (I initially reached out to him with an old fashioned letter), I invited myself to visit so I could find who, if anyone, was developing business applications for the Internet. At that time, there was no better person on earth I could have asked that question of. In the computer industry, when it came to doing practical things with the Internet, Mark was known as "Mr. Internet". In recognition of that fact, he was named one of the 100 most important people in computing by *MicroTimes Magazine* in 1993.

Mark was very helpful and encouraging during that first meeting, as he always was. When we were parting, he told me that a hearing sponsored by the U.S. Department of Commerce was going to be held at the community college campus in Sunnyvale later that week. The topic: "The Business Potential of the Information Superhighway". He didn't have high expectations for the hearing but said that at the last minute, a trade show was added to the program. "You might find some people to talk with there."

A few days later, I joined the commuters and took the train from San Francisco to Sunnyvale where the event was being held. The trade show could not have been more modest. It was on the school's indoor half-court basketball court. There were only about ten tables in total. Card tables.

It looked more like a primary school science fair than anything else. In fact, you'd be hard-pressed these days to find a primary school science fair that looked as basic. But "there was gold in them thar hills" on those card tables.

Infoseek, one of the original search engines had a table. They had a guy there with a PC, a long extension cord, and a card table. The pitch, if I remember correctly, was this: Prepay $99 and get 100 Internet searches. It seemed like a pretty good deal to me. Jerry Yang and David Filo, founders of Jerry and David's Guide to the World Wide Web (now better known as Yahoo!), had only quietly appeared on the Web just a few weeks earlier and were barely known. I certainly had never heard of them. There was no such thing as an Internet search engine then and I figured it would take me a long time before I used up 100 searches.

Slip.Net had a card table too. They were one of the pioneers of selling Internet access to the public in the Bay Area. I hit it off with the sales rep, John Stauber, and for the next year, he and I were in touch regularly. He was the first person I met who was both a salesman and involved with the Internet. Later that year Slip.Net was the primary advertiser in the eight-page tabloid Internet industry news magazine *The Internet Gazette* which I started publishing in the fall of 1994.

Also at the trade show, but not demonstrating, was Bruce Moore of Bernard Hodes Advertising. The agency was the biggest buyer of newspaper ad space in the country. They specialized in job listings for corporate clients and recognized very early that they needed to look online as a way to distribute their product. Bruce told me I should look up someone named Marc Fleischmann.

At the time, Marc was one of the most exotic creatures imaginable: A full-time freelance website developer. I had never met one before and I'm reasonably sure he was the first one in all of human history. There were people who developed websites for fun and people who developed their organization's websites as salaried employees, but making one's entire living just making websites? My first thought was: "How is this poor fellow going to feed himself?" I needn't have worried.

Among other things, Marc was the first person to put the full content of a newspaper on the Web, the *Palo Alto Daily News*. He was also the first person to put a traditional mail-order company on the Web with an e-commerce site that made sense, Hello Direct, a company that then sold headsets for telephone operators and salespeople. Marc did these things way back, long before companies like Yahoo! and Amazon had even registered their domain names.

I came home from the trade show with the names of three people who were making their full-time livings doing business things on the Internet that didn't involve the government, the military, or academia. Given that the only other person I knew before that who fit into this category was Mark Graham, my Rolodex had grown 300% in a day. Things were looking up.

By sheer luck and without planning, when I moved to San Francisco I had landed in the one place on earth where people like Hal Josephson, Mark Graham, and others like them existed in abundance. Worst case, I had to jump on a commuter train for a short day trip. From where I lived, I could easily meet face-to-face with the people who were at the cutting edge of what was then the Internet industry. Had I lived in Topeka, Kansas, or at that time New York City or London, opportunities like this to advance what I was trying to do would have been zero. Geography, as it turned out, is destiny even in the Internet Age.

By the way, what *was* I trying to do?

After reading George Oslin's book about the evolution of the telegraph and Cecil Hoge's book on the history of

America's massive mail-order companies, a third book opened up my thinking considerably not only about what was possible for the Internet but the most likely path its development would take and how I could take part. The book was called *InfoCulture: The Smithsonian Book of Information Age Inventions* by Steven Lubar, now a professor at Brown University. It came out in 1993 so the Web wasn't mentioned. Nevertheless, it could not possibly have been more useful to me.

As Lubar's history made clear, when it comes to media revolutions the same pattern asserts itself over and over again.

First, when world-changing communications technologies are invented, no one, least of all their inventors, has any clue what they are ultimately going to be used for. For example, Marconi was trying to develop a way ships at sea could send and receive telegraph messages without wires, which is why the original name for his invention was a radio-*telegraph*. Alexander Graham Bell thought the best use of his invention the telephone was going to be to pipe live music into people's homes. Thomas Edison was looking for a way to replace the manual relaying of telegraph messages with something mechanical and discovered that the wax cylinders he used to do the job also captured sound. Because of that we now have recorded audio.

Second, captains of industry invariably not only show no interest in new media technologies but more often than not actively ridicule them. Sir William Preece, the Engineer-in-Chief of the British Post Office and a genuine giant of the British telegraph industry, discounted the usefulness of the telephone, then a new invention. "We have a superabundance of messengers, errand boys, and things of that kind." Why use a phone when you can send a messenger who's close at hand?

Third, users invent how new media technologies will be used. Not inventors, not scientists, not manufacturers, not marketers, not the government, not academicians, but the people who actually use the thing. One striking example of this comes from the history of the telegraph. Originally, telegraph machine messages, the dots and dashes of Morse code, were embossed on paper, which were read and transcribed by hand. Then a 14-year-old boy from Kentucky named Tommy Ahren discovered that he could translate Morse code on the fly simply by hearing it. Initially, this skill was considered superhuman, so much so that he was hired by PT Barnum to demonstrate it to audiences as a "wonder of nature." Eventually, it was discovered that Ahren's skill was learnable and all telegraph operators learned to "read" telegraph messages the way he did.

You might ask what does all this have to do with the Internet? From reading Lubar's history it became clear to me that if the Internet were to become something genuinely useful, the engineers would be the *last* to figure out the details. Leaders of legacy industries would have no interest in its potential and even be outright hostile to it. How the Internet would be used would be invented by its *users*, and most probably by people who were not anyone's idea of experts.

From reading all this history, I got the notion that maybe *I* could contribute to what needed to be done and what no one had yet done up to 1994. That would be to figure out how the Internet, with all its promise, was going to pay for itself, the thing that every medium has to figure out. After all, I was perfectly qualified for the job. I was not an engineer and was as far away from understanding the details of the technology as possible. I was also not a captain, or even a corporal, of industry.

What I *was* was someone intensely curious about how this Internet thing was going to pay for itself. It wasn't clear to me and it wasn't clear to anyone that I talked with, but there had to be an answer.

Chapter Thirteen
Meanwhile at Urbana-Champaign

*"The only people for me are the mad ones, the ones who
are mad to live, mad to talk, made to be saved, desirous of
everything at the same time, the ones who never yawn or
say a commonplace thing, but burn, burn, burn..."*

— Jack Kerouac

After months of searching high and low and in every nook
and cranny of the Bay Area, then the undisputed digital
media capital of the universe, I'd finally found a handful of
people who lived on the business side of the Internet. They
were selling Internet access to consumers directly, gateways
to the Internet to companies and website development to
businesses and publications that were testing the waters.

These folks were "all in", full-time burn-your-boats-
on-the-shore types and they had ample motivation and
high hopes that the Internet, specifically the Web, would
succeed as a medium. But hope, of course, is not a strategy
and neither were the occasional spasms of hype that came
from the news media about the Internet and its potential.

They were the pioneers, undertaking the backbreaking
work of clearing land, while not knowing what, if anything,
would successfully grow on it. Assuming they could find
crops that would grow successfully, how they would get
them to market and, once brought to market, who would
buy them were still unanswered questions.

Another unanswered question was the status of Mosaic,
the popular point-and-click web browser that turned the Web

into something usable for everyone. Everything depended on it, but in early 1994, the team that had developed it were college students, recent graduates, or about to graduate, and thus on the verge of being scattered to the four winds.

In essence, Mosaic was a grand piece of hobbyist freeware. "Hobbyist" in the sense it was no one's living, thus no one's job to keep supporting and developing it. "Freeware" in the sense that it was given away for free, and no one had given any thought as to how to receive money in exchange for it.

Having an easy-to-use, point-and-click front end to the Web was then key to the Web having any value at all. Otherwise, accessing the Web was possible, but it was like being on a skating rink without ice skates, or even shoes. Yes, you could get from Point A to Point B, but it wouldn't be very efficient or even comfortable for the average person. With personal computer users accustomed to the Windows or Mac point-and-click interface, no one was going to get excited about having to manually type in commands to make every aspect of the Web work. Plus there were no pictures.

There was talk, and even some action, by companies who were in the early stages of doing something to put their version of a web browser on a commercial footing, but none of it was compelling and what efforts there were were moving in slow motion.

It might be worthwhile to briefly backtrack and take a look at the essential milestones that got us to this point. If you're already familiar with this history, you can skip ahead.

October 29, 1969 - The first ARPANET link was established and the first message was sent (between UCLA and defense contractor Standford Research Institute). By the end of the year, a grand total of four computers were connected to the network. ARPANET was the creation of the U.S. Defense

Department, specifically the Advanced Research Projects Agency (ARPA).

January 1, 1972 - The e-mail format that we all know today was created and added to ARPANET, and within four years, 75% of all ARPANET traffic was e-mail.

January 1, 1983 - ARPANET adopted TCP/IP, a universal computer networking language that allowed all computers, regardless of format, to connect to the network This step laid the technical foundation for the Internet as we know it today.

January 1, 1989 - The National Science Foundation Network (NSFN), which inherited the management of the Internet from ARPANET removed the rules restricting the Internet, as the network was now called, to government and academic institutions only.

November 3, 1989 - CompuServe, the first consumer online service, gave its customers access to the Internet e-mail system. For the first time, it became possible for consumers, in this case CompuServe users, to send and receive messages anywhere on the planet, as long as their correspondent was on a computer connected to the Internet. By the end of 1994, Internet e-mail became ubiquitous for customers of all the big online services and the vast majority of BBSs.

December 1, 1990- Tim Berners-Lee and Robert Cailliau at CERN, the European Organization for Nuclear Research, released the three technologies - HTML, URL, and HTTP - that became the foundation of the World Wide Web.

Cailliau was responsible for a critical but little-appreciated step in the process. By CERN policy, the web protocols that Berners-Lee and Cailliau developed were public domain for anyone to use, but as Cailliau recounts, attorneys representing American institutions like NCSA and others wouldn't accept that fact without CERN issuing a formal declaration of it. CERN's administration was reluctant to bother with the formalities and it was Cailliau who navigated the bureaucratic shoals to make it happen. Had this not been accomplished, it's very possible that the World Wide Web would have ended right there and then.

The first web page was posted later that year and in 1991, users outside of the small world of CERN were invited to use the technology for their own online collaboration and publishing projects. The first web server didn't appear in the U.S. until a year later, in December 1991 (at the Stanford Linear Accelerator Center).

Note how many technical things that were fundamental to the success of the Internet took place in 1989 and 1990, independently without centralized control or a master plan. And they happened off the radar of the news media. The stage was being set, but none of the people who had had a hand in setting it had any idea what was about to happen.

By the end of 1992, there were somewhere between 200 to 500 web servers total, running a few thousand websites. In contrast, today, as I write this, there are now 1,130,000,000 (1.13 billion) websites with approximately 250,000 new ones being added every day.

What the heck happened?

Marc Andreessen and Eric Bina are what happened. Marc was an undergraduate in the computer science program at the University of Illinois at Urbana-Champaign. Eric was a graduate of the same program and was working for the National Center for Supercomputing Applications (NCSA), part of the university. NCSA was founded just seven years earlier in 1986 and was one of the first five supercomputing centers in the U.S.

Champaign-Urbana, Illinois, the small twin city metro area, where the university is located, is smack in the middle of the Corn Belt. The surrounding area is a sea of corn and soybean fields. It's about a two-hour drive south from Chicago, two hours west of Indianapolis, and two and a half hours northeast of St. Louis. Not many distractions and a good place to get work done.

Marc and Eric shared a Siberia-like space in the basement of the university's Oil Chemistry Building (now called the Roger Adams Building) behind NCSA's brand-new five-story brick headquarters. Both were in the habit of working late into the night, Andreessen on a graphics and data visualization program (Polyview) and Bina on a real-time collaborative work program (Collage). They also spent a lot of time talking about the future of computing. One night in November of 1992, Marc asked Eric if he'd ever seen the World Wide Web. He had not. Andreessen showed it to him.

A few weeks later in December, inspired by visions of new features that they wanted to see in a web browser, they started working on a "better mousetrap", a new and improved browser for the Web. Marc focused on the program design and Eric did the programming. The pair worked fast and on January 23, 1993, Andreessen sent this e-mail to the world:

07:21:18-08000 marca@ncsa.uiuc.edu

By the power vested in me by nobody in particular, alpha/best version 0.5 of NCSA's Motif-back networked information systems and World Wide Web browser, X Mosaic, is hereby released:

file:_ /ftp.ncsa.uiuc.edu/Web/xmosaic/
xmosaic-0.5.tar.Z

At the time, the then highly obscure World Wide Web had, despite its grandiose name, less than 1,000 websites on it total. The global number of Internet jockeys may have been relatively small, but they were well connected and influential among computer users in their local communities. Word of an easier and friendlier way to access the Internet spread fast, especially after two new team members created a version for Windows users.

Andreessen, in a scene reminiscent of *The Adventures of Tom Sawyer* and the white picket fence, enrolled a small team of "let's do this just for the heck of it" fellow programmers at Champaign-Urbana and NCSA to convert the program from the original UNIX version that he and Bina had written to versions the rest of us could use.

Aleks Totic handled the Macintosh version. Jon Mittelhause and Chris Wilson handled Windows. HTTP development was the domain of Rob McCool. Cross-platform work with "lots of late-night coffee and beer runs" fell to Chris Houk.

In addition to making the program usable for all personal computer users, the team supported early users, responding personally to their e-mails, answering questions, and providing tech support and troubleshooting. Based on these high-volume interactions, the browser was continuously fine-tuned to make it easier and easier to use.

At some point, executives at NCSA took notice. In an act that will live in bureaucratic infamy, they declared that they owned the program, and started a public relations campaign to "tell the Mosaic story" which, not coincidentally, starred them.

In their biggest PR coup, NCSA reeled in *New York Times* technology writer John Markoff. His December 9, 1993 article "A Free and Simple Link to Computer Network" attributed the browser's existence to NCSA without mentioning the names Marc Andreessen, Eric Bina, or any of the other developers. In the same article, Markoff knew enough to attribute the web's fundamental architecture to Tim Berners-Lee of CERN, but the coffee, beer, and pizza-fueled programmers who conceived the idea of a web browser with a friendly interface and then built, launched, and supported it, were reduced to "a small group of software developers and students".

Amnesia about the creators does not appear to have been an oversight. A month before the *New York Times* article, the globally syndicated TV show *Computer Chronicles* (1983-2002), gave Mosaic its first-ever TV mention in November of 1993. Physicist Larry Smarr, who deserves much praise for helping sell the National Science Foundation on funding the first five supercomputer centers in the U.S., presented the wonders of Mosaic as a product of NCSA. In the process, he failed to mention that the quarterback for the effort was a twenty-two college senior working on his own initiative.[22]

The Mosaic browser was downloaded over 1 million times over the course of its first year of existence in 1993. Relative to today's numbers, one million downloads may not sound that impressive. However, at the time there were less than 15 million people on the Internet total. To achieve a "platinum record" comparable to Mosaic's success in 1993 with over 5 billion people on the Internet today, you'd need about 300 million downloads. Not bad for a new software program with a marketing budget of zero.

---

22   There is a "controversy" over who conceived and underwrote the effort to develop the Mosaic browser. If, as NCSA asserts, they were involved from the beginning then, given the realities of bureaucracy there must be a paper trail showing that Andreessen and Bina were assigned the task by NCSA and were specifically compensated for the hundreds of hours they put into it. If such a record exists, no one has ever produced it.

At the end of that year, Andreessen graduated, left campus (without bothering to pick up his diploma), and joined the rest of the Class of '93 in the search of his first real job. On campus, he'd been paid $6 and change an hour as a work-study student.

A few people on the Internet side of things knew Marc's name and what he'd accomplished, but to Silicon Valley, he was just another bright kid looking for work. One thing he was certain of, thanks to his experience with NCSA, he never wanted to work on a web browser again.

At the very same time, Marc was heading west to California, Jim Clark was walking away from Silicon Graphics, the Silicon Valley superstar company he founded in 1982.

A high school dropout from Plainview, Texas, Clark, received his original training in electronics from the Navy. One thing led to another and despite having never earned a high school diploma, he received a Ph.D. in computer science in 1974 at the age of 30. Five years later, he was an associate professor at Stanford University. Just three years after getting that coveted job, he and some students started Silicon Graphics.

In less than ten years, the company went from some raw academic ideas to being Hollywood's go-to computer workstation and in the process revolutionized movie visual effects and 3-D imaging. Companies like George Lucas' Industrial Light and Magic led the way using Silicon Graphic workstations and software to create never-before-seen movie special effects like those seen in some of the later *Star Wars* movies and *Jurassic Park*. Steve Jobs's company Pixar also favored the Silicon Graphics workstations and used them to make breakthrough animations like *Toy Story* and *A Bug's Life*.

Like Steve Jobs at Apple, Jim Clark eventually found himself frustrated within the company he started. Like Jobs, he felt the CEO, in Silicon Graphics' case Ed McCracken, was taking the company in the wrong direction. Given that the company's market cap collapsed from over $7 billion in 1995 a year after Clark left to $120 million in 2006, he might have been right.

After years of mounting frustration, Clark formally left the company in January 1994, the same month Andreessen moved from Illinois to Silicon Valley. Clark was casting around for a new project. He was convinced, as many were at the time, that interactive TV was going to be the next big thing. As the founder of the company that made the hottest video-processing computer in history, he managed to get meetings with some of the big media corporations that were trying (and failing as it turned out) to make headway with it. Every home would need a box to handle the new technology. Why not a box made and sold by Jim Clark?

It made sense on paper, but there was one problem.

In making his break with Silicon Graphics, Clark had agreed to a non-compete. He was specifically banned from hiring any engineers from Silicon Graphics. Tech companies know that, in addition to investor money, the essential fuel that they run on is their engineers. It's almost easier to find money than it is to assemble a top team. This was a serious problem for Clark's aspirations to start another tech company because the only top gun engineers he knew were his crew at Silicon Graphics. Other companies in Silicon Valley, like Apple and Sun Microsystems, were adept at keeping their golden goose engineers on very short leashes.

What to do? It was not an easily solved problem. Like Andreessen, Clark was all dressed up with no place to go.

In 1994, Jim Clark, the entrepreneurial equivalent of a Formula 1 race car driver, was sitting on the sidelines for lack of access to the high-octane engineers he knew were fundamental to building the kind of business he'd proven himself to be good at building.

Marc Andreessen, the designer/team manager/ promoter of the most successful software program of 1993 - and now arguably one of the most important pieces of software in the history of computing - had just had his creation yanked from him. He had to be wondering what he'd do for an encore and if it was even plausible that he'd ever have an encore.

The community of advocates for Internet technology wanted to connect everyone to the network and had the know-how to do so. They just didn't know what people would actually do once they got there, who was going to create the content for it and why, and where the billions of dollars were going to come from to lay connections to everyone's homes, workplaces, and schools.

Multimedia producers were discovering all sorts of tricks to weave together linked text, audio, and video to put information and entertainment on people's personal computer screens, radically changing the desktop experience. However, whether they realized it or not, the public at large was not and would never be that interested in CD-ROMs and would never pay the prices producers needed to make their business's arithmetic work.

Farsighted advertising agency executives like Rick Boyce who *were* paying attention to the new technologies - multimedia, the Internet, 500 channel cable boxes - had the sense they were seeing something important unfold, but couldn't figure out how it would fit into life as they knew it, namely clients with big budgets writing big checks for media buys.

Then there was me. I wanted the Internet to succeed in the worst way for three reasons: 1) it would give me a new and impactful way to reach my customers without the serious expense of physical media (production and delivery) and I could track the performance of my ad dollars in ways not possible with any other medium. 2) I was an information hound. Bookstores, libraries, and magazine racks were good but not enough, and 3) I wanted to see a self-publishing revolution, one in which everyone who had something useful to share, including me, could make it available to everyone on earth with an Internet connection.[B]

But as the saying goes, if wishes were horses, beggars would ride.

Little did I know that just a few months later in the spring of 1994, two events were going to take place that would push the accelerator to the floor. Those of us who in January 1994 were all dressed up with no place to go were suddenly going to find ourselves very busy. Not only that, we were going to be pulling the rest of the world along with us at a pace that no one, especially not us, could have imagined.

Chapter Fourteen
The First Green Shoots of Spring

"God knows how we'll make money, but OK."
— Jim Clark, co-founder of Netscape

"If the company does well, I do pretty well. If the company doesn't do well, I work at Microsoft."
— Marc Andreessen, co-founder of Netscape

As spring sprung in 1994, I was starting to wonder if the dream of a fast easy-to-use Internet in everyone's home was ever going to be realized. Maybe I was impatient, but the situation was starting to remind me of the play *Waiting for Godot*.

In the play, two men, Vladimir and Estragon, do a lot of talking while they're waiting for a third man, Godot, to arrive. He never does. I don't know what Samuel Becket had in mind when he wrote the thing, but life in San Francisco was beginning to imitate art; there was much talk about the future, no small amount of it delusional, and it was beginning to seem that while we might get a few more bells and whistles on our computers, this Godot fellow was never going to arrive.

Then I read the news. Interestingly enough, it was not in the *New York Times*, the *Wall Street Journal* or even *Wired*. It was in a little print newsletter called the *Internet Business Report* dedicated to Internet issues.[c] Right on the front page. It stopped me in my tracks and made me think, "OK. Maybe this thing has a chance."

The news was that Jim Clark, the accomplished and inspired tech entrepreneur, and Marc Andreessen, who had created the biggest software hit of 1993, had found each other. The two announced the formation of a company called Mosaic Communications and they were going into the software business selling web browsers and servers. The day was April 4, 1994.

*This was the news announcement that got me so excited
I almost fell out of my proverbial chair.*

The alliance didn't come a minute too soon for Marc. While not floundering, he was not exactly thriving in his first job. It was at a Silicon Valley software company called Enterprise Integration Technologies (EIT) that worked on the ARPA-funded initiative CommerceNet. Mosaic colleague Rob McCool said of Marc's time on that first job, "He was just bored stupid. He couldn't stand it." The business climate in Silicon Valley at the time didn't help. Things were so stagnant there that reportedly Marc thought he might need to learn Japanese to advance his career.

It's all obvious now in retrospect but had the two not met, we would probably be living in a completely different world today. Clark had been singing the blues one day to long-time Silicon Graphics colleague Bill Foss. He wanted to start a new company but was prohibited from recruiting all of the great software engineers he knew at Silicon Graphics by the non-compete agreement he signed when he left the company. "What about Marc Andreessen?" Foss asked him. Clark searched his memory and came up blank. "He just moved to Palo Alto from Illinois," Foss added.

Realizing that Clark had no idea what he was talking about, Foss got out his laptop and fired up the Mosaic browser. After a quick demo, he turned the machine over to Clark who was more than a little intrigued. "How do I get in touch with him?" Marc was still using his college e-mail and the address was in the browser notes. Clark immediately sent him a message:

Marc:

You may not know me, but I'm the founder and former chairman of Silicon Graphics. As you may have read in the press lately, I'm leaving SGI. I'm planning to form a new company. I would like to discuss the possibility of your joining me.

Jim Clark

When I read about Clark and Andreessen's new venture I knew nothing of this backstory. It just seemed like a marriage made in heaven: The first person to make a piece of consumer-friendly software for the Web and put it in the hands of over 1 million people joined with someone who had the Silicon Valley experience, connections, and insight to build a commercial foundation beneath the making and selling of the browser and server as a product.

Note that up until this point, the Mosaic browser was freeware built and supported by people who were essentially doing it on a hobbyist basis (even though they were working 12 to 18 hours on it.) which is the exact *opposite* of a business.

Now is a good point to explain the "commercialization" of the Internet. The point of commercializing the Internet was not to plaster over something "pure" and non-commercial with a bunch of advertisements and charge people for things that had previously been free. It was to make the Internet viable and sustainable and, while I'm throwing in buzzwords, diverse. People might be willing to work for free for a while, but eventually, college students graduate and they need to pay rent.

Software needs to be upgraded, bugs need to be dealt with, and support tickets need to be answered. If you aspire to put billions of people on the Web, that means you need salaried employees to do the work to make that happen. This will seem blindingly obvious to many, but we're drifting into a society where ideas like Modern Monetary Theory have some "educated" people believing that all that is needed for a prosperous society is for the government to print money, distribute it widely, and everyone will magically be able to get all the things they want and need.

It doesn't work that way.

To use a mundane example, if you look at a carton of milk on a supermarket shelf, with a little imagination you will soon realize the following: Someone had to pay for that cow, feed it, keep it housed, and take care of its vet bills. The farm the cow lives on had to be purchased. Someone had to build the barn the cows are milked in. Someone made the milking equipment and other people need to maintain it. The cows need to be milked two to three times a day. Someone has to pick up the milk and deliver it to the bottling plant.

Someone else had to manufacture the cartons and deliver to the plant. Then truck drivers need to take the milk from the plant to the supermarket. Someone had to raise the money to have the supermarket built. Someone has to put the milk on the shelves and take the money at the cash register. The supermarket has to be maintained, roof, plumbing, daily cleaning. All the people doing these things need to be paid so they can cover their rent or mortgage, real estate taxes, food, cars to get to and from work, clothes, things for their kids, and countless other personal expenses.

It takes a cast of millions for you to be able to go to the supermarket and get a quart of milk, and they all need to be paid. The process of making milk available is thoroughly commercialized. That's what I mean when I say the *commercialization* of the Internet, operating things in such a way that you can pay the people who do the work it takes to keep the various parts of it running. A volunteer communitarian ethic was not going to cut it in the long run. It takes commercial organization and the discipline that accompanies it.

We can repeat this scenario for every single item on supermarket shelves and then repeat it for any type of store you can imagine. Every material thing we enjoy is the result of things being done in a businesslike way. In short, printing money doesn't create societal wealth. People waking up and doing useful things in an organized way creates societal wealth.

With Jim Clark on board, the chance that quality Web browser and server software was going to be made, upgraded, maintained, and distributed went from zero to "this could actually happen." Amazingly, before this point, the web was based on the largess of individuals who had the vision and did the work and institutions that had resources they were willing to spare. No business, no industry, and certainly no global communications network can be run on that basis.

Thus the overarching importance of Jim Clark.

After several weeks of conversation, Clark and Andreessen concluded that Clark's initial vision of making boxes for interactive television or partnering with Nintendo on a next-generation gaming console was not the way forward. By the process of elimination, the two surprised themselves by coming to the unexpected conclusion that the best idea for their business was to make and market web browsers and servers.

Up until this point, despite the extraordinary popular success of Mosaic, the news media's relentlessly dismissive attitude toward the potential of the Internet while simultaneously rhapsodizing over unworkable business models like interactive TV, actually had Marc discounting the development of software for the Internet as a viable business idea.

With their decision to go into the web software business, the next step was obvious - and urgent. Marc already had an existing team of web software developers who worked brilliantly together back at the University of Illinois. They were a rare commodity and there was a very real danger someone else might snap them up. One, Chris Wilson, already had been. He was signed by a Seattle-based company Spry where he fled to avoid the oppressive work environment at NCSA.

Without wasting any time, Andreessen sent an e-mail to his crew back at Champaign-Urbana with the following message:

> *Something is going down here - be prepared to leave*

As Clark reminisced later, "Marc and I hatched the idea over the weekend, and on Tuesday, we were on our way out there," and booked two tickets from San Jose to Illinois. There was no time to invite Marc's team members to fly individually to California for personal interviews. They had to be scooped up whole and as quickly as possible.

In one day, using a hotel called the University Inn as his base, Clark individually interviewed all the programmers who had worked with Marc: Totic, Mittelhauser, Wilson, McCool, and Houk, whose contributions were listed in the previous chapter, plus Lou Montulli, the guy who created the Lynx browser and later invented HTTP cookies among other innovations. Montulli worked remotely with the Illinois crew from his base at the University of Kansas where he'd been an undergraduate in the computer science program.

Clark was impressed with what he saw and heard and offered everyone a job. While the programming team went out for beers to celebrate their impending release from NCSA servitude, he went back to his room to write up the employment contracts which he sent to the hotel's fax machine to be printed out. Laptops were common in 1994, but hotels that made printers available to guests were not a thing yet. Everyone signed and the next day, they went to the NCSA head office and resigned en masse. Within days, they were all in Mountain View, California in apartments close to the company's original office.

Clark knew exactly what he had to do next. He reached out to John Doerr of the venture capital firm Kleiner Perkins and sold him on the vision. Being allied with the right venture capital firm is a must in Silicon Valley. In addition to the money it provides, an alliance with a good firm brings advice and access to other resources, including all-important human resources. It also provides the credibility needed to attract other potential investors.

Next, Clark brought in Rosanne Siino who he'd worked closely with at Silicon Graphics. She was the person who helped him draft his resignation letter and his message to the media about leaving the company he had founded thirteen years earlier. Siino was critical to the new company's success. Creating brilliant software is one thing. Selling it is another and the selling begins with making sure that people know the company exists. If it was going to succeed, the new company was going to have to develop many, many relationships fast and the way was paved by Siino who knew how to shape, tell, and spread the story.

Today he story is obvious: Wiz kid and Silicon Valley veteran team up to create a company that transforms the Internet. It was not an easy story to sell *before* it happened. Siino not on y had to introduce the new company and Marc to the world, she also had to bring people up to speed on what the Web was, what a browser was, and why any of it mattered. Time was of the essence.

The plan to release a commercial browser and server software was guaranteed to put the new company on Microsoft's radar as a threat to their aggressively managed monopoly control of the desktop. Microsoft's Info Superhighway plan, which was already in the works, involved a set-top box co-designed with John Malone's Tele-Communications Inc. (TCI). It would serve as a front-end to a new generation of digital televisions the content of which Malone, his colleagues in the cable business, and presumably Bill Gates, would control.

Fortunately, it took Bill Gates over a year from the founding of Mosaic (later renamed Netscape) to publicly acknowledge the Web as a threat to his enterprise and by then it was too late. The World Wide Web had escaped the barn and there was no bringing it back, but Clark, Andreessen, and Siino didn't know that at the time. So the pressure was on.

As I said, when I read the original company formation announcement, I knew none of this backstory. What I did know was that the combination of Clark and Andreessen had tremendous promise. I looked up Marc Andreessen's e-mail address, I wrote him, and he wrote back. Later I learned that Marc was a champion e-mail correspondent. In those days anyone who wrote him with a halfway intelligent question about Mosaic was assured of an answer.

I didn't have technical questions. I had a business question. I reasoned that ultimately it was content, not software, that was going to bring "regular" people to the Internet. Who did he imagine was going to pay to have this content made?

We'd both been following a high-profile story about the CEO of Procter & Gamble complaining about the effectiveness of the money his company was spending on advertising. P&G, which makes soap and other consumer packaged goods that fill whole aisles of the supermarket, has consistently been the #1 advertiser in the world for decades (though in recent years Amazon has dislodged it from time to time). When P&G complains about the performance of the ad agencies it hires, which I learned noticed is a cyclical event, Madison Avenue quakes in its boots.

Marc and I talked about that for a while. It was clear to me, and I believe to him too, that at the end of the day, advertising money of some kind was going to be necessary to support content. After all, newspapers, magazines, radio, and TV all depend on advertising for their survival. However, among many die-hard early netizens, the idea of ads on the Internet was anathema.

Marc, who knew that culture well, posited that broad-appeal web content might be sponsored by corporate advertisers. Interestingly, P&G had pioneered this approach on radio and television with the invention of soap operas. They were called "soap" operas because they were produced by P&G as a way to gather a mass audience to market their soap and related products to. It was a potential model for web advertisers, but I didn't think it was going to be the answer.

I must have made a good impression on Marc. Later in the year, when he passed on making an appearance at the annual COMDEX show in Las Vegas, he, with the encouragement of Rosanne Siino, made the time to drive up from Mountain View to speak at an event I hosted in San Francisco in November. But I'm getting ahead of myself.

Around the same time that I'd reached out to Marc, I'd noticed, thanks to my frequent readings from George Oslin's history of the telegraph and Steve Lubar's history of media technology, that the 150th anniversary of the first official demonstration of the telegraph was coming up soon. When the telegraph was born on May 24, 1844, John Tyler was president of the U.S., the country's tenth. The nation's population was less than 20 million, Chicago had 4,000, and the first California Gold Rush that transformed San Francisco was five years in the future.

I'm a big believer in one-time San Francisco denizen Mark Twain's dictum "History doesn't repeat itself, but it often rhymes" and I thought, what better way to mark the telegraph's 150th anniversary than a meeting about the Internet and its future? I'm also a big believer that if you host a meeting with the clear intention of getting something specific done, more often than not, it will get done. In my naivete, I decided this would be a perfect occasion to bring the smartest Internet people I knew together to sort out, once and for all, how the heck the Internet was going to pay for itself.

For speakers, I asked Mark Graham, already an Internet commercialization veteran; Marc Fleischmann, the world's first full-time website developer; and Rick Boyce, a media buyer from the ad agency industry who aspired to figure out how advertising would find a place on the Internet. They all said yes.

For the meeting, I rented the penthouse conference room at 3220 Sacramento Street. In addition to the meeting room and a small kitchen, it had a big deck overlooking the city. It was the perfect place for a party or a small conference.

3220 Sacramento Street, the place where so much happened. Now it's the South Campus of San Francisco University High School.

I'd learned about the space from Mark Graham whose company Pandora had offices there. Its formal name was 3220 Gallery, but everyone called it 3220 and it had a most auspicious history. Among other things, Apple Computers had run a multimedia lab there where its engineers worked on things like touchscreen technology and QuickTime. For a while, QuickTime was the dominant format for digital video, and of course, years later, the touchscreen technology they worked on there became a very big deal.

The first commercial e-mail link that connected the U.S. with the former Soviet Union, a citizen's initiative, was housed at 3220. It served clients like the *New York Times*, Stanford, Harvard, and the Library of Congress at a time when communicating with the Soviet Union was a tenuous thing. During the 1991 coup attempt, there were days when this was the only communications link to the Soviet Union from the outside. 3220's owner, Henry Dakin, not only provided the service's founder Joe Schatz with subsidized office space, but he also wrote him a $5,000 check for seed money on the spot after hearing his pitch. An act of this kind by Henry was not a one-off event.

If anyone had a promising idea related to promoting peaceful relations between the U.S. and the Soviet Union, reducing the chances of nuclear war, citizen diplomacy, education, protecting the environment, harnessing new technologies to expand communications options, parapsychology, or new concepts in physics, Henry might offer them seed money and free or subsidized office space, furniture, and a T1 Internet connection. This was back at a time when fast connections to the Internet were exceedingly rare and expensive. He supported so many projects that ten years after his death in 2010, a 464-page book *Henry S. Dakin: From Physics to Metaphysics & Guns to Toys* was published. It's filled with short narratives from some of the people and organizations whose work he helped jump-start.[D]

I didn't know any of this history when I booked the meeting space for my small Internet get-together. It just looked like a nice place to have a meeting and it was within walking distance of my apartment. Henry was interested in what I was doing and offered me the space for a nominal amount. The space had already been rented for the meeting day I wanted, May 24, 1994, the 150th anniversary of the first official demonstration of the telegraph, so we held it on the first available Saturday which was June 11, 1994.

I had a simple goal for the meeting.

I was going to bring together the three smartest people I knew who were interested in the question of how to put the Internet on a sound commercial footing. I invited an audience. Together, once and for all, speakers and audience were going to figure out a model for advertising on the Internet that made economic sense and would provide a financial foundation for online publishing ventures large and small.

Chapter Fifteen
Childhood's End

"Every creative act is a sudden cessation of stupidity."

— Edwin Land, inventor of the Polaroid camera

Pre-Internet, research meant a physical trip to the library or a bookstore to see if there were books and magazines on the subject you wanted to know more about. If the problem was big enough and you had sufficient resources to hire someone to handle it for you, the Yellow Pages might work.

These sources might point you to academic courses, consultants, and organizations with conferences and specialty print directories and newsletters which your library almost certainly would not have. Today, of course, in many cases, you can find answers and resources within minutes on almost every subject under the sun from how to fix a leaky sink to getting a handle on calculus.

But Internet or no Internet, what do you do if you're searching for the answer to a problem that no one had solved yet? This was the situation every person interested in the financial future of the Internet faced in the spring of 1994.

As 1994 dawned the challenge was this: Yes, the Web was exciting in principle, but how was it going to attract the broad public and how was it going to support itself financially? The Internet needed not just technology enthusiasts and journalists but also moms, school kids, buyers for industrial concerns, young adults trying to find their way in life, and all the myriad of people who needed to show up for the Internet to really work. Without them, the

Web would never raise itself above the level of a glorified BBS.

To my simple mind, the answer was advertising. How did newspapers and magazines pay for themselves? Advertising. How did radio pay for itself? Advertising. How did TV pay for itself? Advertising.

Was the Internet likely to be any different? I didn't think so, but many of the people who were building the Internet and many early users were vehemently opposed to advertising on the Internet of any kind. They not only didn't want to see direct response type ads, they believed commercial activity of any kind could only harm the Internet.

Objections to advertising on the Internet came from people like Tim Berners-Lee, the co-inventor of the Web, and Vinton Cerf the "father of the Internet" and co-developer of the core protocols (TCP/IP) that the Internet ran on. Critiques of Internet commercialization appeared on CNET and in the pages of *Wired* (ironic, right?) Academics, journalists, non-profits like the Electronic Freedom Foundation, and thousands of "netizens" joined the fray on the "anti" side. Note that what the vast majority of these people had in common was that they'd never run a business, and never had to make payroll or pay a server bill in their lives.

Their concerns broke down into four issues:

1. Erosion of the Internet's original purpose as a free and open resource for everyone
2. Cluttering
3. Surveillance and privacy
4. The emergence (or rather re-emergence) of digital "walled gardens" in which a few corporations would control people's access to the online world and the content on it

To address these one by one:

1. For those who make even the slightest effort to learn Internet business basics, advertising has not interfered with the aspiration of regular people to share information online. On the contrary, it's made millions of online publishing ventures possible that could never have existed without it.

2. Yes, some websites are cluttered with ads. These kinds of websites are easily avoided and in the event they have information a user wants, scrolling past the ads works just fine.

3. Surveillance and privacy is a genuine concern and there are websites, like Facebook and others, that have surveillance and accompanying violations of privacy as part of their business model. Related to this are companies like Google which, along with Facebook, engage in censorship and propaganda programs on behalf of advertising clients, government agencies, and political parties. Despicable in the extreme, but this has nothing to do with people running ads to support their businesses. My personal solution to this problem is to boycott companies that engage in this behavior.

4. A few digital "walled gardens" have most definitely emerged, with Facebook being the most egregious example, but their walls all have gates and the gates are open. I consider the fact that so many people think that Facebook *is* the Internet and rarely, if ever, roam from it when they're online, a colossal waste of human potential. However, pre-Internet, there were many people who despite mobility and literacy never left the region they grew up in, never went to a library, and never read a book. The existence of advertising does not produce people like this and sometimes advertising is the only way to help people find worlds beyond the one they were born into.

It was a nice theory that somehow the Internet could keep itself running and that people would spontaneously create a superabundance of varied and worthwhile content for it without some kind of sustainable financial underpinning. But until someone comes up with a better answer than advertising, advertising is it. ("Advertising" would include web publishers who sell subscriptions, special access, and products and services of interest to their audiences.)

Advertising was the answer, but it was also the question. What *form* would Internet advertising take? No one knew, but Rick Boyce, a media buyer for the advertising agency Hal Riney, had given me a very important clue when I visited his office and he showed me the volumes of the Standard Rate and Data Service. That's when it became clear to me that whatever form Internet advertising took it had to be something simple, straightforward, and *standardized*. Media buyers needed to be able to buy ads without having to reinvent the wheel with every new order.

At this point, in May 1994, I didn't know many people who'd be interested in this topic, but I did know three really good candidates: Mark Graham, who'd been helping people with their Internet-based projects for years; Marc Fleischmann who had been building commercial websites for clients almost as soon as the first Mosaic browser was available; and Rick Boyce, who knew the media buying side of the advertising world as an insider.

Surely I thought, if the four of us got together and focused on the question of what form advertising on the Internet would take, we had a good chance of figuring it out in a day. On the one hand, my optimism was ludicrous. On the other hand, by this point I'd been to dozens of panels and talks on "the future of media" and I had yet to see one case that focused on this particular problem. What people were focused on, and rightfully so, was selling whatever they were selling, and in 1994 that meant Internet accounts for consumers and Internet-related technical services to companies.

Asking busy people to get together and brainstorm all day was not going to work, so I structured it in the form of a seminar. Seminars are great business tools as I learned from the mortgage industry conference business that I'd built and was still running at the time. They can be profit centers all by themselves. It's a relatively painless way to create content (just push "record"). Hosting a good one is a way to publicize your expertise and authority in a field and to build your professional network fast. It's a way to create new customers and create closer ties with existing ones. And if you do it well, they can be fun (though the host doesn't necessarily have much fun until the after-party).

Most importantly to me, seminars are a peerless way to get people out of their normal routines and give them a chance to open their minds and think.

I'd learned from producing concerts, something I'd done in my teens through early twenties, that event production is easy. You only need four things: a time, a place, an act, and an audience. I had the first three: sometime near the 150th anniversary of the telegraph, Henry Dakin's penthouse conference space, and the four experts.

To get an audience I wrote a sales letter that featured a snapshot of Steve Jobs, Steve Wozniak, and my friend Dan Kottke manning the primitive Apple booth at Jim Warren's Computer Faire in 1977. (For perspective, at the time, Jobs was 22 years old and Wozniak was 26.) The theme of my letter was that if you missed the personal computer revolution, "now you're about to get a second chance."

These two subheads appeared on the first page: "The personal computer is on the brink of a second major earth-shaking, cash-spewing revolution - and this time non-techies are going to share in the bonanza" and "Imagine: marketing minus postage, printing, or fulfillment costs". Where I had gotten my certainty from, I have no idea. Fifteen people showed up. Enough for a quorum. By the way, I was

If you had bought these guys a cup of coffee, you might very well be a multi-millionaire today

And now you're about to get a second chance.

Believe it or not, you are looking at the founders of a Fortune 100 company. They started it in a garage with capital raised from the sale of a broken down VW Bus. Less than seven short years later, they calculated their net worths in the hundreds of millions of dollars each.

(Thanks to my friend Dan Kottke for the photo. Dan's on the far right. The other two guys are Steve Jobs and Steve Wozniak, the founders of Apple Computer. Here they are at their first trade show. They were so green at the time they didn't even think to have a banner made with the company's name on it.)

Great story, you say, but what does this have to do with me?

A lot. Listen carefully. . .

The personal computer is on the brink of a second major earth-shaking, cash-spewing revolution - and this time non-techies are going to share in the bonanza

If you only understand one thing about computers, understand this:

Unlike most things in this world, computers are getting faster, and better, and cheaper all the time - and there is no end in sight.

For example, in the next two years Sony plans to come out with a game player, a kind of mini-personal computer, that will have the same processing power as the first Cray supercomputer which sold for a cool $10,000,000. Sony "guesstimates" their version will cost about 500 bucks.

And technology experts say that this kind of mindboggling progress is going to continue month after month, year after year right into the 21st century (which in case you didn't notice is less than 75 months away.)

Imagine: marketing *minus* postage, printing, or fulfillment costs

If you're an advertiser you know all about postage, printing, and fulfillment costs. They eat up a major portion of your company's revenues. I know because they eat up a major portion of mine.

You know the routine. You run an ad. Prospects call your 800 number (cost). The calls are answered (cost) and the addresses transcribed (cost). Then you send off a brochure, sales letter, or catalog which always costs too much to print (cost) and too much to mail (cost).

The sales letter I wrote to attract people to the June 11, 1994 seminar.

174

perfectly happy to have beginners at this meeting. First, almost everyone then in the Internet space was a beginner, and second, smart beginners are often good at asking fundamental questions, and sometimes in the process of answering them, you can get new insights.

The day was broken up into six parts.

9 AM to 10 AM - I spoke on *"Absolute Online Basics for Absolute Beginners"* as a pre-seminar session for people who didn't know much. This was something I was eminently qualified to do given I'd spent the last ten months as an absolute beginner. I defined terms like computer (I kid you not), software, modem, baud and BPS, BBS, online service, and Internet. In my handout, I included these statements: "Current (modem) speeds will look like a joke in a few years" and "Audio and video are technological possibilities for the not-so-distant future".[23]

10 AM to 11 AM - I gave a talk called *"Modems, Media, Marketing, and Money"* in which I laid out the current state-of-the-art of business online. This talk, given on June 11, 1994, is as close to a perfect time capsule of conditions as they were up to that day. How can I say that? Because I had two genuine in-the-trenches practical experts on the subject in the audience - Mark Graham and Marc Fleischmann. Had I gotten something wrong, neither would have been shy about letting me know. I waited for corrections and they didn't come. I also laid out the history of how direct response advertising fueled the development of all media - print, radio, and television - as well as summarized the principles of direct marketing practice.

11 AM to Noon - Mark Graham explained the Internet. By that point, he'd been doing high-level things on it for well over ten years. I looked forward to Mark's talk because I still had a lot of basic questions myself and I figured the

---

23   If you call the eleven years it took for Internet video to be commonly available to consumers (YouTube in 2005) as the 'not so distant future', I was right on the money.

audience would ask most of them for me. (Another value of holding seminars.) Mark had more experience with down-to-earth applications of the Internet than just about anyone on the planet at the time, and he was well practiced in explaining the Internet and, by necessity, had made scores, if not hundreds, of Intro to the Internet talks.

Noon to 1:00 PM - If you ever put on a live event, food and drink matter. I had a secret weapon in this department. Bettina made soup and sandwiches. As a former restaurant owner and cook, it was child's play for her, and I'm pretty sure we were in the top 1% of all catered business meetings worldwide that year. Nothing fancy. Just good solid homemade food that everyone appreciated. (Her cookbook *A Taste of Heaven and Earth* is available on Amazon.)

1:00 to 2:00 PM - I asked Marc Fleischmann to talk on Multimedia on the Internet, a topic of deep fascination to me. He dispensed with it rather quickly: "There is none yet." Then he talked about his real-world, here-and-now projects. Marc's focus was on building websites that sold things.

At that time, Marc had few peers in this arena. For perspective, the founding of Amazon was two months away. Within six months of the original Mosaic browser appearing on the Internet (the summer of 1993), he built what may have been the very first commercial third-party web presence that combined the Web, search, and ordering into a single system. It was called the Document Center and it quickly doubled his client's revenue. And he didn't do this by just putting up a few web pages. He handled every part of the task from configuring the web server to writing the custom programming that was needed to get the site to do the things he wanted it to do.

As far back as 1993, he was counseling clients that the number of hits to a website was meaningless, that flashy graphics didn't necessarily help and could actually hurt sales, and that being named the "cool site of the day" had zero value. It was years before the vast majority of people building websites for a living caught up with his 1993 observations. Some still haven't.

Marc had the perfect background to be one of the world's first business-minded web marketing consultants. His father had owned an advertising agency and Marc started doing graphic design and layout as a teenager. No computers then. Just drawing tables, X-acto knives, tweezers, and rubber cement. To give you an idea of how prescient he was, in an interview I did with him in 1995, he pointed out the following: "Driving people from existing media to the online catalog and getting them to call on the phone and order (yields) the highest return with the least investment." This is yet another thing he pioneered that few businesses have caught on to and the few that do think they invented this approach in the 2000s.

The seminar was a success. It started and ended on time (don't laugh, people care about things like this), it was held in a pleasant environment, the food was good. The speakers were top-drawer and shared information you'd be very hard-pressed to find anywhere else all in one place.

But there was a problem.

One of my missions for the meeting was to give Rick Boyce something concrete he could take back to his office and put to use as a media buyer. On that score, I felt the meeting had failed miserably. If you had come to the meeting wanting to know how to build an online catalog that operated profitably and made sense, Marc Fleischmann had laid it all out perfectly. However, if you were in the advertising agency business and your job was creating and placing ads, we hadn't shed any useful light on that at all. In my defense, when this meeting was held there was no online ad industry. The world had yet to see its first banner ad. That would take place six months later. Yahoo!, which for a time was the portal to the vast majority of web users, wouldn't run its first ad until January 1, 1996, our meeting was on June 11, 1994.

As the meeting ended and we were wrapping up, an idea popped into my head. I asked Mark Graham if he'd sit down with me and Rick. I needed Mark to make sure that what I was about to say was technically possible. Up to that point, I'd never run a web server, built a website, or made a single web page. My idea was pretty simple and it was born out of desperation to give Rick some value from the meeting.

I had intuited that if the Web was going to succeed, and I really wanted it to, it was going to have to be supported by advertising just like every other media. Rick was not only the only person I knew who was open to thinking about these things, but he was also the one and only person I knew in the mainstream advertising industry. How many meetings could I reasonably ask him to attend? How many speculative conversations could we have without some concrete actionable suggestions flowing from them? My guess was not many. He was here now. I better come up with something - and this is what I said:

"Rick. Here's something you can do. You can put a little square on a web page and then when people click on it, it takes them to a big page that you control where you can tell your whole product story without limit. Not only that, you'll know how many people saw the ad and you can count how many people clicked on it so you'll know how effective your ad was." I was so green about Internet technology at the time, that I had to ask Mark if the latter part of my suggestion, the ability to record page views vs. clicks, was doable. He said it was.

I didn't realize it at the time and neither did Mark or Rick, but in less than 60 seconds I had just laid out the foundation for Internet advertising, which became the financial underpinning of the World Wide Web.

And then things started to really heat up.

Chapter Sixteen
The Big Show

"There's something about taking a plow and breaking new ground. It gives you energy."

– Ken Kesey

Unbeknownst to me or Rick at the time, two months before the June 11 seminar at 3220 another meeting was taking place that was going to have a huge impact on the Web. It involved Andrew Anker and *Wired.*

Anker had spent years covering the media, technology, and communications industries for banks like First Boston and Sterling Payot. He'd taken time off to work at a start-up that was based on the cable TV model of the information superhighway. It didn't pan out, but it gave him insight into the potential of what might be coming. When he first saw the plans for *Wired,* the magazine, in the early 1990s, he was smitten and became an investment fundraiser for it. His family put the first $25,000 into the round he managed at a time when the founders were struggling to find money.

The magazine which, thanks largely to Anker's fundraising efforts, launched January 1993 at Mac World was an instant success. Previous to *Wired,* all publications related to tech were narrow industry journals - tech industry news, new product announcements, and ads for hardware and software. Back then, computer people were considered "back office" workers, out of sight and out of mind. A few people like Steve Jobs and Steve Wozniak had made it to public awareness, but they were the rare exception.

Tech folks responded very positively to seeing themselves portrayed in the pages of *Wired* as innovators, visionaries, and heroes - people who mattered. *Wired* was a perfect publication-to-audience match and went on to become the go-to source for all global media outlets who were looking for interesting tech stories. The match was so perfect that unlike every other consumer publication in the modern history of magazine publishing that I'm aware of, it launched without the need for a massive direct mail campaign to scale its subscriber base.[24]

In March of 1994, consulting with *Wired's* publishers, Anker developed the business plan for Hotwired.com, an Internet-based publication. Like the print magazine, it would cover the intersection of tech and society with a focus on personalities and lifestyle. The plan was approved in April, Anker was named CEO, and that summer they started building it.

Hotwired.com represented the first time in history that a business put resources into developing a brand-new website with the intention that it would be supported by selling advertising. But before we talk further about Hotwired.com, we need to mention Global Net Navigator (GNN), which was the first website to sell ads to support itself. In August of 1993, a year before work started on Hotwired.com, Dale Dougherty, Lisa Gansky, and web designer Jennifer Rogers launched Global Net Navigator (GNN).[E] The effort was supported by O'Reilly Media, a publishing company which Dougherty co-founded with Tim O'Reilly, former tech writer. O'Reilly Media was making bank selling Ed Krol's 1992 book *The Whole Internet User's Guide and Catalog*.[25]

---

24   Richard Benson's *The Secrets of Direct Mail* goes into depth about all that is involved in successfully launching a new magazine. He was the launch quarterback for magazines like *Architectural Digest*, *Psychology Today*, *Smithsonian*, *New York Magazine*, and many others. They all required heavy upfront investment in direct mail promotions to build a base of paid subscribers. *Wired* didn't.

25   *The Whole Internet User's Guide and Catalog* had already sold over 250,000 copies by 1994.

The original intention behind GNN was to provide a directory of web links, essentially the same model that the founders of Yahoo! started working on five months later. At the time GNN launched, websites were rare and hard to find. Simply presenting "all" the world's websites in an organized directory was a major innovation. Revenue was an afterthought and it wasn't until early 1994 that they started selling link sponsorships on their site. Basically, "write a check, get a link to your homepage." Not the most sophisticated form of Internet advertising, but a step closer.

Meanwhile, I was working on my next project. The June 11, 1994 seminar at 3220 in San Francisco had been a success. The speakers enjoyed themselves and the attendees were happy. The experience inspired me to attempt a more ambitious event and I started work on it right away.

Hard as it may be to imagine now, but as late as the fall of 1994, there had never been a conference anywhere on earth, not even in "digerati" heaven San Francisco, that focused specifically on how to do business on the Web. Meetings among individuals planning specific ventures, yes. Public gatherings no. At the time, it didn't look like anyone was going to put one on so I decided that I would.

Using my formula of time, place, "show", and audience, I decided to focus on the show first. Who would be on the podium? Mark Graham and Marc Fleischmann of course. I wanted Rick Boyce to be a speaker to give attendees his view of the Web from the point of view of someone in the advertising industry. When I asked Mark, Marc, and Rick not only were they in, but they suggested other presenters, and in a short time I had the foundation of a solid program. I continued to reach out to more people, but I didn't have a "name" speaker.

One of my three guiding principles of business is: "No matter what business you think you're in, you're in show business." And the one thing every business and conference can benefit from is affiliation with someone who has name recognition. Marc Andreessen was not a name to the public yet and surprisingly few people in the multimedia world knew his name, but the Internet people sure did and I knew they were going to make up a big part of the audience for the event I had in mind, so I set my sights on getting Marc as a speaker.

Luckily, it was not a cold call. I'd previously reached out to Marc after I'd learned he'd gone into business with Jim Clark and I'd had a few intelligent e-mail exchanges with him. Marc knew who I was,[F] but whether he knew who I was or not, I had to have something more to offer him than "Hey, a bunch of us are getting together to talk about business on the Web. Do you want to come?" That was not going to work.

Fortunately, I had something that Marc and Mosaic Communications (soon to be Netscape) needed that would help them advance their mission: a strong connection to the Bay Area's multimedia industry which I could uniquely deliver as a receptive audience. Years later, I learned that before their spectacular 1995 IPO, Netscape was having serious challenges getting media industry people to take them and the Web seriously.

Clark and Andreessen knocked on a lot of doors and while there were some exceptions, most companies, including ones that mattered like Knight-Ridder, the massive newspaper chain, the *New York Times*, and even Time Warner, didn't see how the Web's then-small audience merited their attention let alone serious investment. Their very reasonable question was "How are we going to make money with this thing?" In 1994, there was no answer to that question.[26]

---

26   One noticeable exception was William Randolph Hearst III, who was so intrigued by the Web's potential that he left his position as the editor and publisher of the *San Francisco Chronicle* and joined Kleiner Perkins Caufield and Byers, the same Silicon Valley venture capital firm that was the first to invest in Netscape.

When I told Marc I could deliver an audience of Bay Area-based multimedia producers, ad agency execs, Internet people, and people from corporate communications departments, he was intrigued and put me in touch with Rosanne Siino who handled Mosaic's public face to work out the details with me.

Pursuing my mission to build a big audience for the event, my first call was to Hal Josephson. I'd written ad copy for his cutting-edge multimedia show-and-tell conferences in the early 1990's. He'd been head of the *International Interactive Communications Society* (IICS). They had chapters in Los Angeles, San Francisco, Seattle, Dallas, Chicago, Minneapolis, Atlanta, San Diego and Detroit. At its peak, the IICS had over 3,200 dues-paying members. Hal was no longer leading it, so he referred me to the new president Jeannine Parker who responded to the idea of a conference about the business on the Web with maximum enthusiasm.

Jeannine told me she'd promote news of the event to IICS's e-mail list. Targeted e-mail lists like this were fairly rare at the time and her list included not only dues-paying members but also thousands of people who were interested in interactive communications. She connected me with Maurice Welsh who was Pacific Bell's New Media marketing director and he contributed a beautiful meeting space at their corporate headquarters at 140 New Montgomery Street. It could comfortably hold 125 people and it was available on Saturday, November 5, 1994. I took it.

Now having a time, a place, and a reasonable chance at a decent audience to make the event worth Andreessen's while, I pitched Roseanne Siino on why Marc should carve out a Saturday to join us. I must have been persuasive. She said yes and Marc was booked. That was July 1994.

As the months progressed, interest in the World Wide Web in general and Marc in particular started to grow logarithmically. Siino succeeded in selling Fortune Magazine on the idea of adding Mosaic (Netscape) to their "The Top 25 Cool Companies for Products, Ideas, and Investments" feature. The byline was July 11, 1994. The article provided the following stats for the company: "Sales: None, Employees: 17".

Had I made my pitch for Marc to be a speaker a few weeks or even days later than I did, I'm sure the answer to my request would have been no. I had barely squeezed through the gate before it closed. As we got closer to the November 5 date of my seminar, I learned that Marc had declined to make an appearance at COMDEX which was scheduled that same month. When I heard that news I was a little worried.

In 1994, COMDEX was the largest and most important computer trade show in the world. Bill Gates thought it was important enough to be there and gave the keynote that year. His topic was "The Information Superhighway and the Next Era". As we got closer to the date of my event, I was getting not-so-subtle messages that Marc might not be able to make it to my event after all. Yikes.

In short, the good news was I had a rising new superstar on the lineup. The bad news was I had a rising new superstar on the lineup and he might not show.

Meanwhile, I kept looking for ways to build the audience. Sometime in the summer of 1994, I met Jim Warren (1936-2021) who was the keynote speaker at ONE BBSCON '93. An absolutely fascinating man worthy of his own book.[G] One day, I took Jim to lunch in my neighborhood and he told me the story of how he promoted his successful and historic Computer Faire.[27]

---

27   Steve Jobs and Steve Wozniak had a booth at the inaugural Faire in 1977. They were so green at the time that they didn't think to bring a sign for their new company Apple, so they defaulted to using the sign the trade show had made to mark their exhibit booth. The "exhibit booth" was a card table with a curtain behind it.

To promote the Faire, Jim created an eight-page newspaper, printed on newsprint, which he called the *Silicon Gulch Gazette*. He packed it with interesting stories each one ending with a pointer to the Faire, and devoted the last four pages to info about the three-day event including a registration form.

"Wasn't that expensive?" I asked.

"Not at all. Newsprint is the cheapest form of printing there is."

There used to be an old saying, now out of favor in our entitled era: "Take a millionaire to lunch." Well, I did and walked away with an idea conservatively worth 1000x the cost of Jim's lunch. When I got back home, which was conveniently located just across the street from the restaurant, I started to work on my own eight-page newspaper *The Internet Gazette and Multimedia Review*.

Internet Gazette and Multimedia Review
October/November 1994, Volume 1 Number 1

It seemed a bit odd, even to me, to start a *newspaper* about the *Internet*, but in the summer of 1994, there was no place to buy ads for an Internet conference. The arithmetic was splendid. I was able to print 25,000 copies for $1,000 or 4 cents a copy and I got that from the sponsor of the back page, the Internet service provider Slip.Net, that I'd met in Sunnyvale earlier in the year.

The first issue came out in October 1994 and it was designed to promote my upcoming November 5 conference. I had settled on a name for the conference: *Multimedia Publishing on the Internet - Opportunities for Publishers, Advertisers, and Entrepreneurs*. The front page of the *Gazette* carried two stories, one about video on the Internet and one about what we anticipated the Internet would do to the music industry entitled "The Music Industry is Dead. Long Live Music!" I was so new to the newspaper publishing game that it didn't occur to me to put, or even write, headlines for these stories and put them above the fold.

The *Gazette* included a calendar of local events which included listings like: October 1 - *Getting Online for Women*, October 2 - *Internet 101*, October 8 - *Internet for Activists*, October 8 - *SF PC Users Group*, and so on. I wrote a column called Net Business Briefs in which I compiled Internet-related news items. This first column included the sale of the Internet Shopping Network to Home Shopping Network and *Wired* hiring Rick Boyce away from Hal Riney & Partners for Hotwired.com which I called "its new online marketing venture".

There were three more issues after that. In the second issue, I declared that we were in the business of training, design, production, technical consulting, marketing, and publicity services for companies that wanted to do something on the Internet. I also announced the existence of a then-imaginary Web School with four tracks of classes – Intro Track, Programming Track, Publishing Track, and Business Track - to see what, if any, the demand would be.

In the third and fourth issues I ran a full-page ad with the ominous - and now prophetic - headline "Could your company be banned from the Internet?" I was advertising our domain name research, consulting, and registration services. This ad proved to be a big winner. It included a "Free Internet Domain Name Search Form" which invited businesses to submit the domain name they'd like to use and two alternates in case the one they wanted was taken.[28]

Now back to the November 5, 1994 conference. For the first time in human history, people were going to gather specifically to focus on the practical issues of how to use the World Wide Web to do business and make money by providing content.[H]

With the enthusiastic and generous promotional support of Jeannine Parker and the *International Interactive Communications Society*, the numerous friends I'd made in the local multimedia and Internet industries, *The Internet Gazette*, and the fast-rising star of Marc Andreessen, the event sold out. At the very last minute, *Wired* submitted the names of six of their reporters for press credentials. I didn't need them, but *noblesse oblige* style, I figured out how to squeeze six more chairs into the room.

All was going well but for one thing. The folks at Mosaic, which right then was in the process of changing its name to Netscape, started asking me to remind them who

---

28   Before 1999 when the first domain name registrars were licensed, Network Solutions Inc. had a monopoly on all domain name registrations, and like all monopolies, they were not easy to deal with. In 1994, very few businesses had the first clue about how to look up let alone register a domain name so we made it easy for them and handled all the hassles involved. It became the perfect front end to getting consulting clients and selling web design and hosting.

As of 2023, global domain name registration is a $2.3 billion a year industry and growing. As far as I know, we were the first people to create a specific service around researching and registering domain names. Companies like San Ash Music, Boardroom Books, and many others acquired their domain names using our services during this era.

I was and what this meeting was all about. This was not a good sign. Between July and November, Marc had become a mini rock star and was well along the road to becoming a big one.

I wasn't paying Marc anything to appear and we had no contract, just a "Yes. He will be there" from corporate communications. They didn't come out and say he wasn't coming, but I could tell that if I didn't energetically re-sell the vision, chances were good we would not see him. When I learned he'd blown off COMDEX the same month because he was so busy, I got the proverbial sinking feeling in the pit of my stomach.

My pitch was simple: The whole digital interactive multimedia industry of San Fransisco was behind the event, everyone who was anyone in the industry was coming, and we'd built the whole event around Marc and Mosaic. The last part was 100% true. The first two parts were poetic license with a reasonably strong dash of truth.

When the day arrived, the room filled early and despite the gloom and heavy rain that morning, there was a festive atmosphere. It felt more like a party than a business conference. The people in the room had clearly been looking forward to the day. They were especially excited about finally laying eyes on Marc Andreessen.

And I still had no idea if he was coming or not.

When the clock struck 9 AM (or maybe it was 10 AM) I led off. (The talk I gave that day is in the back of this book, Appendix L.) I'd hoped Marc was going to hear it, but there was no sign of him. Next up was Mark Graham who laid out Internet basics, something that was still necessary to do then even in San Francisco and in front of a technically sophisticated audience. And still no Marc Andreessen.

I went back to the control booth to listen to Mark Graham's talk. (Yes, there was a control booth in the back of the conference space. This was Pac Bell's headquarters after all.) My talk had gone well, Mark Graham's was going well and I knew we were going to have a highly worthwhile panel of experienced online business people: John Barnhill of Silicon Reef; Ann Covey of The Well; Marc Fleischmann, pioneer web presence producer; and Bruce Moore of Career Mosaic, a division of Bernard Hodes, the world's biggest buyer of newspaper ads.

But at this point, I was focused on how I was going to explain that the conference's headline act and the person that many, if not all, had come to see, was not going to make it. Then, halfway through Mark Graham's talk, a very tall young guy in khakis and a white shirt walked through the door stage right of the dais.

"Is that Marc Andreessen?" I could not be sure because of the distance and the fact I'd only seen pictures of him. I ran over and it was indeed him. I took him back to the control booth and we watched the remainder of Mark's talk. Two things stood out: he crackled with intelligence and he had a good sense of humor. He'd also mastered some of the sound effects from *The Simpsons* especially Homer Simpson's "D'oh!"

He showed up.
(Left to right: Ken McCarthy, Marc Andreessen, Mark Graham)

When it was Marc's turn to speak, he used transparencies and an overhead projector. No PowerPoint or unknown, and therefore suspect, Internet connection for him - even at Pac Bell's corporate headquarters. From giving dozens of presentations on the road, he knew better.

Marc failing to show proper deference to his elders

Marc started by going through the process he and his team at the University of Illinois used to develop the first Mosaic browser and how the still-new company replicated that method, though not the code, for their first product which had launched a little more than two weeks before on October 13, 1994.

Their method: move fast, do the easy stuff, leave the hard stuff for someone else, ship, and don't agonize. Nine days later, on November 14, the company changed its name to Netscape, and by March its new product had taken 75% of the existing market for browsers.

In addition to being first with the best and most stable browser, Marc & Co. pioneered two things now common in software development.

First, "rolling beta" testing. Beta testing, which first appeared as a concept at IBM in the 1950s, meant testing a full feature version in contrast with alpha testing which meant testing product ideas and theories. Beta tests were originally conducted in-house, but expanded to include select invited users, usually from members of the company's trusted customer base.

Netscape pushed the concept of "beta test" to the max. Back at the University of Illinois, they made beta versions of Mosaic available to anyone from the general public who wanted to download it and try it out. This meant quick and abundant feedback which the Mosaic team had trained itself to respond to with equal speed leading to faster development cycles and a better product. This was software development at a speed that had never been seen before. Ten years later Mark Zuckerberg summed up the "rolling beta" ethic which had become the norm for companies like Facebook and Google as, "move fast and break things."(Zuckerberg left out the "and fix them fast" part.)

The second contribution is so basic that it was easy to overlook. Previous to Netscape, software was sold in physical form, on floppy discs, CDs, and later DVDs, and came in shrink-wrapped boxes. You bought the box at a store or you ordered it via mail order. (Before that, software came embedded in the hardware it ran on. The two were once sold as a single inseparable unit.) Netscape pioneered and normalized the now-standard process of commercial software being distributed online. Today, you download software. You don't drive across town for it or wait for the box to come in the mail.

My favorite part of Marc's presentation was the question and answer session and my favorite exchange went something like this:

> *"Marc. I was driving on Route 101 and I saw a billboard with a URL on it. What do you think?"*

(This may well have been the first time a web address appeared on a billboard anywhere. Until the fall of 1994, it was not something people had thought to do.)

Marc's playful answer which had the audience roaring with laughter: *"Yes, it's pretty bizarre. It's pretty strange. They're actually taking it seriously."*

The tone of his voice indicated that even though he'd been working around the clock to get this software to the world, even he was shocked at the speed with which it was now suddenly being embraced.

The link to the complete video of Marc's talk is on the Online Archive page at the back of this book. As far as I know, and I've been looking for thirty years, we are apparently the only people from that time who had the presence of mind to push the "record" button when Marc was pitching the browser and the Web as the gateway tool for the future.

DEC 28 '94 03:53PM NETSCAPE COMMUNICATIONS 415 2542   415-563-348.272

Speaker Release Form

1. In exchange for promotional consideration, I am appearing as a guest speaker at an event sponsored by E-Media of San Francisco.

2. I understand that my presentation will be audio and videotaped and, in exchange for promotional consideration, I am granting E-Media complete license to distribute these tapes for the purposes of promoting both my own company and E-Media.

3. I further understand that in some cases these tapes will be distributed for free. In other cases, they will be distributed for a fee.

Name _MARC ANDREESSEN_   Date _12/23/94_

Company _Netscape Communications_

Title _VP Technology_

Marc Andreessen's signed speaker release form.

The poster for the Multimedia Publishing on the Internet conference. The tiles on the left are from Venice. The Chinese characters that appears on the right-hand side say: Hau lang tui qian lang ("Things develop ceaselessly with the new superseding the old")

The ad for the Multimedia Publishing on the Internet conference, as it appeared in the Internet Gazette. My friend Ornette Coleman was performing the following week, so we put in an ad for his show. Pictures of Marc Andreessen were hard to come by at the time and our designer, who was working from a group photo of the original Mosaic programming team, mistook Jon Mittelhauser for Marc.

- Part Four -
Now What?

Chapter Seventeen
Have You Ever? You Will.

"The future is already here - it's just not evenly distributed."

- William Gibson

One person was missing from our November 5 meeting: Rick Boyce.

Hotwired.com had launched nine days earlier on October 27, 1994. It marked the first time a website went into business with the intention of creating, not scraping, original content[29] and selling ads to pay for the enterprise. Rick led the team that sold the ads and they succeeded in completely selling out all the ad positions in Hotwired.com's first issue.

When we last saw Rick it was June 11, 1994, and he was a media buyer for Hal Riney & Partners. Months earlier, I'd been the first person to explain what the Web was to him. I suggested that it might eclipse the 500-channel cable box vision that was all the rage then and put us in a media world unlike anything we could ever imagine. He was interested and open-minded. Then a few months later, at the June meeting that I described in Chapter 15, I suggested to him that one possible model for Internet advertising would be to "put a little square on a page, people can click on it, it will take them to a big page, and advertisers can count and

---

29   Amazingly, as late as 1994 and well into the 1990s, there was still a debate over what was going to be "king" on the Internet, cool technology or content. One camp asserted that somehow an ever-changing progression of "animated dancing bears" and other ephemeral technological gizmos would be enough to attract people to the Web, engage them, and give them a reason to come back for more of the same. I disagreed.

see how many people saw the page and how many people clicked on it."

What I was describing was a basic two-step sales sequence. Nothing new to anyone steeped in direct response advertising and direct marketing, but new to the Internet world. The function of the first ad is to get people to inquire. The function of the second ad is to lead people to buy. Sales sequences can be longer and more involved than two steps, but this is multi-step selling at its most basic.

Brand advertisers, which are the overwhelming majority of clients for advertising agencies, don't think this way. Generally, their job is to remind people that when they go to the supermarket, department store, drug store, or car lot, that the advertiser's product is there, it's great, and they should buy it. This might work for items that are already on store shelves, but that's a very narrow band of the full spectrum of things that are sold in this world.[30]

In contrast, many products and services need extensive explanation before a prospect is ready to buy, thus print or broadcast ads that invite people to write in for a free report or call a phone number for a free consultation. The first ad gets the prospect in the door. The second ad starts the sales process.

This is what I had in mind when I described my model to Rick, but it wouldn't be until four years later in February of 1998 when Bill Gross introduced the concept at the TED conference[1] and put it into practice under the business name GoTo.com. More about that later.

---

30   In truth, advertising agencies that serve big brands don't give much, if any, thought to closing sales with consumers. The main sale they focus on closing is the selling of their services to big brands. The people who are their customers are Chief Marketing Officers whose fuzzy goals often boil down to a vague notion of "brand awareness". Professional success among brand-focused advertising agencies is measured in industry awards for best creative and best campaign and holding on to a client for as long as possible until they get restless and try another bright and shiny object.

Global Network Navigator got the ad ball rolling by selling links to websites, but this was not Internet advertising in the sense we think of it today. Hotwired.com began by selling corporate sponsorships, the model Marc Andreessen had suggested to me in the spring of 1994. However, instead of selling links as GNN did, Hotwired.com sold "little squares that when you click on them take you to a bigger page and you can count the number of clicks."

Many histories of this era call Hotwired.com's ads the very first banner ads and it's claimed by various sources that the first banner ad was by AT&T. The copy read "Have you ever clicked your mouse right here? You will", which was based on a remarkably prescient television campaign AT&T rolled out in 1993:

TV spot #1

Have you ever borrowed a book from thousands of miles away? You will.
Driven across the country without stopping for directions? You will.
Or send someone a fax - from the beach? You will.
And the company that will bring it to you?
AT&T.

TV spot #2

Have you ever paid a toll without slowing down? You will.
Bought concert tickets from cash machines? You will.
And the company that will bring it to you?
AT&T.

TV spot #3

Have you ever watched a movie you wanted to the minute you wanted to? You will.
Have learned special things from faraway places? You will.

*Or tucked your baby in from a phone booth? You
will.
And the company that will bring it to you?
AT&T.*

TV spot #4

*Have you ever kept an eye on your home when
you're not at home? You will.
Or gotten a phone call on your wrist? You will.
And the company that will bring it to you?
AT&T.*

TV spot #5

*Have you ever opened doors with the sound of
your voice? You will.
Carried your medical history in your wallet? You
will.
Or attended a meeting in your bare feet?
And the company that will bring it to you?
AT&T.*

TV spot #6

*Have you ever had a classmate who is thousands
of miles away? You will.
Conducted business in a language you don't
understand? You will.
Or attended a meeting in your bare feet?
And the company that will bring it to you?
AT&T.*[31]

I don't know what's more science fiction-worthy, that
we have all these technologies today, or that AT&T portrayed
them in a television commercial campaign in 1993, years and
sometimes decades before they were technically possible.[J]

---

31   One place you can view this ad television campaign is here:
https://bit.ly/ATT-spot

In truth, the AT&T banner first appeared three weeks earlier on Global Net Navigator. The company that made the ad was Modem Media.

Modem Media was founded by G.M. O'Connell and Doug Ahlers in Westport, Connecticut. The company opened its doors on October 19, 1987, the day of the big stock market crash. That was the same day I was standing on Fifth Avenue half a block from my then-employer First Boston's offices trying to convince a cotton trader friend who'd taken an extended break from the market and thus hadn't seen the day's news, that the market had really dropped 500 points before lunch.

When they started Modem Media in 1987, O'Connell and Ahlers were in their twenties. Ahlers was the tech guy and O'Connell was the sales guy. For perspective, at the time, there were no laptops, there were no graphics, and text on a computer screen was green, amber, or white on a black background. Modems operated at 300 baud.[K]

Many of the deep-pocketed sponsors that lined up to run ads when HotWired.com launched did not have websites and none of those who did had sales pages on them. The whole thing was a brand-building exercise for all of them. They wanted – make that needed - to be seen as "being in on the new thing", the Internet, and to show how cool they were. HotWired.com was the best organized to take their money (with our colleague Rick Boyce running the checkout stand.) AT&T was roped into the adventure because their arch-competitor MCI was part of the launch. The other sponsors were Volvo, Club Med, Sprint, IBM, and Zima (a Coors carbonated alcoholic beverage that thankfully is no longer marketed anywhere on Earth.)

Clicking on the AT&T banner took people to a website that featured images and information about seven great art museums around the world. In those days, companies were not sure if it was acceptable to sell directly on the

Internet so there was much talk about giving consumers "the opportunity to interact with the brand", whatever that meant.

What did it cost to put a banner ad on Hotwired.com? A page in *Wired* cost $10,000. Based on this someone came up with the idea to charge Hotwired.com's advertisers $10,000 per month for a three-month minimum to sponsor a content section, for a total ad buy of $30,000, (though I also heard talk of a three-month "beta trial" for $15,000.) These ad buys were the size of rounding errors for the annual ad spend of these advertisers, but they were eye-opening to everyone who was looking for a model to make web content pay for itself.

AT&T's banner ad was developed by Modem Media, but the destination site - the online showcase of museums around the world - was built by Adam Curry's New York-based OnRamp, an early website developer for corporate clients. (This was the same Adam Curry who was a VJ on MTV, registered the domain mtv.com much to the consternation of his employer, encouraged Halsey Minor to register the domain name cnet.com for what was then a cable TV channel idea, and is considered the godfather of podcasting.)

Hotwired.com made no promises as to how many people would see the ad, but they did measure click-throughs and the initial rates were insanely high by today's standards. These early ads had 40%, 50%, 60% and more click-throughs. That meant approximately one out of every two people who had the ad displayed on their screen clicked on it. Click-through rates for banner ads vary, but today a 0.4% click-through rate is considered respectable. Why the 99% decline in response? When the earliest banner ads ran, there was an extreme shortage of things for web surfers to click on. Early web users were click-happy.

There were no ad servers then. The ads had to be hard-coded onto the pages. There were no analytics programs. The server logs were checked by hand to see how many times a page was displayed and how many people clicked on the ad, the percentage being the click-through rate (the metric I'd introduced Rick Boyce to nearly five months earlier). There was no targeting or split testing though advertisers could change their ads once a week.

In short, Hotwired.com launched commercial web publishing with all the technical sophistication of a high school newspaper, but it was a very important start. As Goethe (1749-1832) said, "Whatever you can do or dream you can, begin it. Boldness has genius, power, and magic in it." And indeed there was magic in it.

This same basic format funds content on the Internet to this day. Selling banner ads (they are now called display ads) is expected to be a $174 billion business in 2024, the year I'm writing this. (Source: Statistica.com)

The search engines Lycos and Excite followed HotWired.com's lead though inexplicably it took them 6 or 7 months to get started. Yahoo! didn't begin running banner ads until January 1, 1996, more than a year later. There were still no ad servers and ad placements were managed with spreadsheets and placed by hand, one at a time.

According to Reuters, a company called Webtrack listed the top five online ad spenders of the 4th quarter of 1995 as AT&T - $567,000, Netscape - $556,000, the Internet Shopping Network- $329,000, NECX Direct (a computer reseller) - $322,000, and Mastercard - $278,000. If these numbers seem low, they are. I have colleagues today, many of them individual entrepreneurs, who spend multiples of these amounts in a single month.

With the advent of the banner ad, the commercial Internet finally had two legs to stand on and it was off and running.

After my November 5, 1994 talk (the text is in the back of this book, Appendix II), I spent much of 1995 guiding San Francisco's multimedia developers on how to make the transition from being multimedia title producers to being website developers for clients. I know of at least three multimedia shops that made the switch the same day they heard my talk. Multimedia developers who were struggling to make their numbers work by developing CD-ROM titles found website creation to be a much better business.

In 1993, four years later, the Fisher Center for Real Estate and Urban Economics at the University of California at Berkeley released a paper by Kenneth Rosen and Avani Patel on the economic impact of what they were still calling the "multimedia industry" on San Francisco. Of the 12 largest "multimedia companies" in the city, 10 were directly involved with the World Wide Web, either making software tools for it or creating websites for corporate clients.

I had my own web consulting, developing, and hosting business which I called E-Media. I not only owned the domain name, I trademarked the term and in 2000 someone made me an offer for the name that was too good to refuse. Ownership of the name and some other intellectual property related to video got sold and resold and eventually ended up in the hands of Akamai, a NASDAQ company.

In early 1996, someone hired me to do research on the Internet market in Japan. The guy who commissioned the research never paid for it or picked it up but by lucky happenstance a few days later I was invited to be part of a program at the Japan Society of Northern California. The U.S. was the first Internet market to take off and it took some time for other countries to catch up. In 1996 Japan was where the U.S. had been in 1994, right on the launch pad.

Ken McCarthy
President

2130 Fillmore Street
Suite 288
San Francisco, CA 94115
Ph: 415 928-4072
Fax: 415 563-3402
ken@e-media.com
www.e-media.com

e-media·

There was e-mail, so why not e-media? I periodically spent money on an attorney to give notice to companies, including IBM, that e-media was a registered trademark.

I received an astonishingly good, even reverential, introduction from the host of the event, went to the podium, and let loose with a stream of up-to-date statistics on Japan's Internet use and growth. Because I'd just done the research, I had no need for notes. Little did I know, NEC's Vice President of Multimedia was in the audience. (NEC is the Japanese equivalent of IBM.) He hired a fellow named Shinji Takeuchi to track me down and check me out and the next thing I knew I was down in San Jose having a very Japanese one-on-one meeting in an empty conference room. With the chief.

The neighborhood where I lived bordered Japantown with its wonderful Kinokuniya Bookstore and I'd read several books on Japanese business etiquette and doing business in Japan. When the big boss handed me his card, I examined it closely and then placed it respectfully in front of me for the course of the conversation. (This is how you do things in Japan.)

The guy was intimidating. When I answered his questions, I watched his reactions very carefully. As long as he smiled, I kept talking. When he stopped smiling, I stopped. Ten minutes into the conversation which was more a visceral exchange than an intellectual one, he signaled Shinji to get rid of me.

Shortly afterward, much to my surprise, I was offered a not-terrible contract to write a weekly report on the San Francisco Internet industry for an NEC project called Biglobe which became one of the biggest Internet service providers in Japan. From 1996 to 2001, the time I wrote Biglobe's weekly Internet industry column, I was "famous in Japan". I never visited the country so I never cashed in on it. Japantown was the closest I ever got.

However, not long after NEC signed me with all the resulting publicity, a Japanese book packager named Hiroshi Kagawa came calling and asked if he could interview me for five two-hour sessions for a one-time flat fee of $10,000, no royalties. With my blessing, he turned the interviews into a book with the English words INTERNET BUSINESS MANUAL on the cover and my name in Japanese.

I don't read Japanese so I have no idea what the rest of the cover said or what was in the book. It was the first book on doing business on the Internet published in the Japanese language. I have a feeling it sold well because Hiroshi went from being an independent book publisher to owning his own publishing company shortly afterward.

In 1997, I had some fun experimenting with a format that later became known as a news blog. Using a diary format, I was exposing facts about the blatant election fraud surrounding the 1997 stadium bond referendum in San Francisco. The proposition was to make a free no-strings-attached gift of $250 million to Edward DeBartolo, Jr., owner of the 49ers football team. The mayor at the time and the chief supporter of the deal was Willie Brown.

ケン・マッカーシー 著
賀川 洋 編・訳

The book that made me even more "famous in Japan". The cover featured the "welcoming cat" ("maneki neko" in Japanese). This lucky charm often appears at the entryway to Japanese restaurants. The raised left paw is meant to attract customers to the business.

Brown is the guy who recruited, trained, mentored, and propelled Diane Feinstein and Nancy Pelosi into politics. When I was battling Brown and his dodgy stadium "deal" with my blog, his then-current crop of promising young protégés included Gavin Newsom and Kamala Harris. Many of the people involved with the 49er stadium campaign ended up indicted and jailed but on charges unrelated to the 49er stadium election fraud.

Eddie DeBartolo, Jr. himself got pinched by the FBI when he gave former Louisiana governor Edwin Edwards a briefcase filled with $400,000 in cash in exchange for a riverboat gambling license. Thanks to DeBartolo getting himself arrested, the bond deal never happened. My blog had so much detailed evidence of the fraud that even Wall Street underwriters couldn't avoid the stench. The taxpayers of San Francisco were saved half a billion dollars, $250 million for the bond, and $250 million over 30 years in interest.[32]

The official history of the Internet says that the first blog to cover a news story appeared in 1998 with the *Charlotte Observer's* coverage of Hurricane Bonnie. I'm sorry, but 1997, preceded 1998 so it appears my blog was the first. As late as 1999 there were only 23 blogs. By 2006, there were nearly 50 million.

In the 2000s, I applied some of the same guerrilla techniques I'd used to defeat Brown to help stop the building of what would have been North America's second-largest cement plant on the banks of the Hudson River, a project of the Schmidheiny family, one of the richest in Switzerland. The family is known to have collaborated with the Nazis during World War II when their company

32   Hilariously, when Carmen Policy, who had been tasked to babysit Eddie, Jr. by his father Edward Debartolo Sr., quit as CEO of the 49ers to go back to Cleveland to be CEO of the Browns, local San Francisco newspapers "discovered" that previous to coming to San Francisco, Policy been the "number one mob attorney between New York and Chicago." Somehow they'd managed to miss this fact for the over ten years he'd been in San Francisco.

operated asbestos and cement plants for the regime using slave labor. Their proposed Hudson Valley plant would have devastated the region and turned it into an industrial sacrifice zone. Instead, the Hudson Valley remains one of the most picturesque regions in America and has evolved into an East Coast version of California's Napa Valley but featuring locally-grown food, restaurants, and locally-made beer, cider, and whiskey instead of wine.

I also assisted Sandy Rosenthal in a project to demonstrate that the great New Orleans flood of 2005 which killed over 1,200 people and created over $100 billion in property damage was not the result of Hurricane Katrina, but rather of the systemic failure of the U.S. Army Corps of Engineers designed and built levee system. Even more impressive than getting to the truth was Sandy's successful battle to compel news outlets that got it wrong - the *BBC*, the *New York Times, AP*, and the rest of them - to correct their shoddy reporting. The story, and my role in it, is covered in her book *Words Whispered in Water*.

But back to San Francisco and back to the 1990s...

Despite some successes, there was one thing that frustrated me. I was glad to have contributed the idea of the click-through as the key metric model for monetizing the Internet. Also satisfying was helping shepherd San Francisco's multimedia industry from the dead-end road it was on to the lucrative business of building websites.[L] It was neat to be among the first to get practical Internet business advice to Japan, then the second largest economy in the world, just as they were starting *their* Internet boom. It was also fun to thwart at least one of the schemes of one of America's most corrupt politicians.

All that was fine, but when it came to the one thing that mattered to me the most I had no luck at all. And that was finding an audience for my insight that the Internet was a direct marketing medium. It appeared to be a fruitless

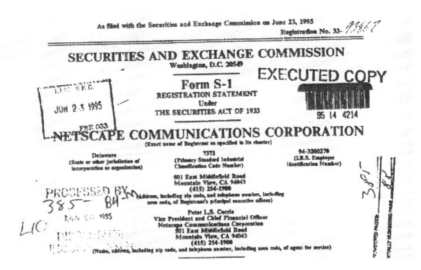

The cover of Netscape's SEC registration for its 1995 IPO. These are known as "tombstones" in the financial industry. On the one hand, this document signaled the birth of the dot-com boom, but it also signaled the introduction of a new element into the freewheeling, share-and-share-alike ethic of the early years of the Internet.

mission. In the early 1990s, people's eyes just glazed over at the mention of the topic and it became exponentially worse after Netscape's IPO, on August 9, 1995, which was just nine months after the November 5th seminar.

Originally, Netscape's stock was going to be offered at $14 per share, but it soon became clear, from the number of people who called the company directly asking for shares, that demand for the stock was going to be off the hook. (Rosanne Siino had to set up an entirely separate phone center to politely deal with the hordes who called in seeking shares, to keep them from overwhelming the company's phone system.) At the last minute, the stock's underwriters Morgan Stanley doubled the offering price to $28. On opening day demand was such that at its peak the share price reached nearly triple that amount. By the end of the year, shares were trading at $174, more than 10x the original proposed price of $14.

Wall Street paid attention. The news media paid attention. The Internet graduated from "cool" to a financial "sure thing". Overnight seemingly unlimited amounts of money became available for Internet ventures.

The concept was simple: If a company driven by twenty-something-year-olds and less than two years old that had yet to show a profit could go from first $14 to $28 to $174 in less than six months, then anything Internet-related could theoretically do the same.

Suddenly, anyone who could write a coherent business plan or (fantasy) related to the Internet and shop it around could get $5 to $10 million. A new metric emerged for these businesses: "the burn rate". The issue wasn't how much you were selling, or how many customers you had, or what your profits were but how much money you were losing each month in a mad dash to achieve some kind of critical mass, real or imagined, that would allow you to "do a Netscape".

It struck me as utter madness, but no one close to it could see it. The mantra of the day was "The Internet changes everything." It sure did. Underwriting standards for initial public offerings went into the toilet.

After the players received their checks, the first thing they did with their money - after throwing lavish parties and fitting out their offices with top-of-the-line Aeron chairs and Sub-Zero refrigerators - was to hire PR firms. Then they'd start buying advertising including expensive television ads.

If they happened to be in the business of selling banner ads, they had a unique vehicle for quickly inflating the value of their companies. It was simplicity itself. For example, Company A would buy $1 million in banner ads from Company B, and then Company B would take that $1 million and buy $1 million in banner ads from Company A. Through this method both companies saw $1 million in ad sales. Wash, rinse, and repeat all across the Internet "ecosystem" and add zeros. No one played this game better than Steve Case and AOL. His ability to acquire Time Warner in exchange for AOL stock marked the height of this madness. I commented on it in writing the day it was announced, but the *New York Times* and *Wall Street Journal* cheered it as the herald of a new age.

In this climate, no one was bothering with anything so mundane as selling products and creating paying customers. Businesses were in search of higher visitor counts so that they could sell (really exchange) more ads and boost their revenue and thus their imaginary value. Profits from the operation would come later. Or not. Netscape had shown the way.[33] You could get Wall Street funding with less than two years in business with no profit. If this sounds like madness, it was.

If people in the digital world were not interested in direct response before the Netscape IPO, they *really* had no interest afterward. The idea of generating revenue by selling things to customers was replaced by the "burn rate", the mad dash to an IPO, the next round of financing, or acquisition by a company that wanted to bulk up quickly for its own financing merry-go-round or IPO.

It was bound to end badly. And it did.

---

33   Unlike many, if not most, of the dot-com entries of the mid-late '90s, Netscape *was* geared to sell product and made a valiant effort to sustain itself that way. Once Microsoft roused itself from its somnambulant state related to the Internet, it fired up its tried-and-true predation machine and crushed the upstart company using its monopoly over the personal computer operating system. Both the U.S. Department of Justice and European Commission found that Microsoft engaged in anti-competitive practices related to the browser market. However, as with many things legal, the judgement was too little and too late to be of any good to the aggrieved party.

Chapter Eighteen
The After Party

"Where are the customers' yachts?"

– Fred Schwed
Stockbroker turned author

My original plan for this book was simple. I was going to collect transcripts of some of the talks I gave and articles I wrote in 1994 about likely paths of evolution of the Internet, put them in historical context with a short introduction, and package the thing as a book for prospective and current clients and students.

Obviously, it didn't turn out that way. Fate intervened.

Bettina and I live in a comfortable old house in a quiet village in the Hudson Valley. We've been here since 1998. We're on a near traffic-less street, but thanks to the local college, we're just a block and a half away from six restaurants, two small food stores/sandwich shops, and an ice cream place so good that some people drive from an hour away just to get an ice cream cone. The village post office and library are on the same "downtown" street. No one planned it, but it's non-urban planning at its finest.

It's four-season weather here and we get a lot of nice days, but in the summer we can have stretches of miserable summer heat exacerbated by the fact the Hudson Valley is in a valley, so it can get very humid. We'd been handling it with some window air conditioners. I didn't think we had a choice in the matter because there's no way to put central air into a house like this after the fact. (This house is so old there are hand-forged nails in some of the floors and I imagine the boards were milled at one of the local water-powered sawmills that used to dot the area.)

Then one of my nephews, an architect, visited us with his girlfriend during a particularly hot stretch and said: "You know Uncle Ken, you could put in a small-duct high-velocity HVAC system and have the equivalent of central air." It sounded like a good idea and Bettina called around and found someone to do the job. One of the machines that's part of the system has to be in the attic and when the guy came to look things over, he told us "I can do the job, but you're going to have to clean up that attic to make room for the equipment and all the hoses and ducts."

The attic. Gulp!

I do a fairly good job of keeping my library, current papers, and the course materials I've created pretty well organized, but everything else - including my papers from San Francisco in the 1990s and New York and Princeton before that - not so much.

When we moved from San Francisco to our home here, I just emptied my file cabinets into banker's boxes. They ended up in the attic as does everything else not immediately connected to something I'm working on at the moment that doesn't fit into my file cabinets. Articles (my own and others), programs and posters from conferences, correspondence, and photographs all got "archived" in the following way: 1) throw it in a banker's box, 2) when the banker's box is full, put it in the attic. Twenty-five years worth of this along with the twenty-plus years of stuff before that. Since some of it goes back to grammar school and since I'm 65 at the time of this writing, it amounts to over 50 years of accumulation.

This "archive", along with receipts, invoices, business records, tax records (some going back before 2000!), and other normal domestic stuff and the attic was packed. Bettina and I had a lot of work to do and other than a few things I knew that I'd saved, like copies of *The Internet Gazette* from 1994 to 1995, I didn't have a good sense of what was up there. Out of sight, out of mind.

We ended up throwing out eighteen large garbage bags, and twenty-two cartons and banker's boxes full of papers. A guy came and shredded all the financial stuff. Clothes that hadn't been worn in years went to the village thrift store and bit by bit, we brought some order to the chaos and carved out room for the HVAC installation.

As I was going through the mess, I was delighted to discover that some of the treasures of my life I had assumed were long gone were safe and sound. I had one folder mysteriously marked "TC", which I later remembered meant "treasured correspondence". (No one should ever hire me to be an archivist.) In it were letters from David Ogilvy, Arthur C. Clarke, Eugene Schwartz, Marty Edelston, Ed McLean (serious direct response people will recognize the latter three names), Ed Niehaus, Rick Boyce, Hal Josephson - and all my e-mail correspondence with Marc Andreessen and others at Netscape during the summer of 1994 printed out. (Hey, maybe I'm not such a bad archivist after all.)

Finding these letters made me think there must be other gems hidden away in the 10 or so banker's boxes we did save, so I set up a folding chair and a coffee table in the attic, and in the heat of the summer, I spent a ridiculous number of hours going through everything item by item. Most of it was added to the trash, but I ended up with a banker's box of "artifacts" from 1993-1995 that boggled my mind. I'd been busier than I remembered[34] and I realized that I'd saved a lot of paper artifacts from a period in San Francisco's history that were not only important to me personally, but could be of interest to historians or history-minded people who wanted to know the details of how the Web won.

---

34   In researching this book, I discovered that "What do I do if I can't remember my employment history?" is a popular search phrase. Before I wrote this book, I would definitely have had trouble with that one.

Most of the existing books and articles about the Internet of the 1990s focus on the successes and excesses of that era. There are countless accounts of success stories like Netscape, Yahoo!, Amazon, and AOL, companies that came out of nowhere and in the space of a few years changed daily life in ways that would have been unimaginable even as late as 1993. Excesses like those fueled by Wall Street fraud and enabled by the financial news media have been well documented too. There have been memoirs of people recalling their experiences with specific companies they worked for, but the cultural ferment (I've always wanted to use that phrase in a sentence) that existed in the pre-boom era (1993-1994) and the *work* that so many people did to lay the foundation for what we have today has largely been overlooked and is in real danger of being lost to history forever.[35]

Even the much-vaunted ChatGPT (as of October 2024) is clueless about much of this history and an extraordinary percentage of the time when I went to it to refresh my memory about a name or a company it confidently returned answers that were completely wrong. I might not always remember a name or a date, but I know when a name or date is wrong.

Some examples:

When I asked for the name of the software engineer who did so much to improve video encoding and who died on September 11, it came back with Michael "Mickey" T. Pohl. Completely wrong. It was Danny Lewin, one of the founders of Akamai.

---

[35]   As I write this (October 22, 2024), the Internet Archive, which has stored over 99 petabytes of data, from old web pages from famous sites like Yahoo! to websites that disappeared long ago, has gone offline. Their flagship service is called the Wayback Machine, and the narrative is that it's imploding from a combination of hacker attacks and vulnerable infrastructure. Mark Graham, who has appeared at many critical junctures in the Internet's history, and my own, is currently the Wayback Machine's technical director. (In case you're curious, one petabyte equals one million gigabytes.)

When I asked for the names of the partners who founded Modem Media, the pioneering online advertising agency, it came back with G. Michael "Mike" Donahue and Gerald "Jerry" Michaliski. Completely wrong. It was GM O'Connell and Doug Simon.

When I asked for updates on the career of Chip Hall, who was one of the people in the audience at the November 5, 1994 seminar we did with Marc Andreessen, it completely missed the fact that among other things he was one of the key people at Google who brought programmatic advertising to the ad buying marketplace. In the history of Internet advertising, this would be like failing to mention a basketball player who had been an MVP in the playoffs.

I could list a few dozen more astonishing lapses in accuracy, but ChatGPT has been frozen all day, and nothing I've been able to do has restored things so that I can review my logs. Suffice it to say, that anyone who uses ChatGPT to try to trace the early history of the Web is going to be sent down many wrong paths. The system does post this warning at the bottom of every page: "ChatGPT can make mistakes. Check important info." This is a bit like posting a warning message on a car: "The car may or may not be able to get you to the supermarket and back."[M]

I realized that much of the early history of the Internet's commercialization that I took for granted was not common knowledge, even among scholars and journalists who write about it as part of their profession. Thus, in my early planning for this book, I decided to add a second chapter filled with interesting factoids drawn from my banker's box of treasures. That was many chapters ago.

Every time I thought I'd be able to sum up the whole story in the next few thousand words, I realized there was one more chapter I had to write so as not to leave out important background material that might otherwise disappear into the mists of time. The more I wrote, and recollected, the more

I marveled at all the parallel initiatives that were going on independently of each other in those especially fertile years of 1993 and 1994.

At a certain point, I asked myself "If I don't tell this story, who will? Who else *can* tell this story?" Understandably, everyone was working in their individual silos. The Netscape folks were racing to get their software up and running. The Hotwired.com folks were racing to get their content and ad sales ramped up. The Yahoo! folks, all two of them in a trailer, were hard at work on a task that ultimately became impossible, hand collating a directory of all the world's websites. The multimedia people were working on their titles. Journalists were filing the stories assigned by their editors on whatever the meme of the week happened to be, missing much, including *Wired* failing to mention Marc Andreessen until October of 1994. Equally as astonishingly, in its premier issue the magazine failed to use the word "Internet" once.

I started thinking about all the professors who are teaching "History of Technology" and "History of Media" courses. There are a lot of them. How can anyone teach this story without addressing the history of the Internet and the critical formative years when it transformed itself from an academic-military-only network that banned commercial messages into the ad-supported Internet that we have today? This history is not just useful to history students. It also has important lessons for students studying business, economics, engineering, and social sciences.

Then I started thinking about my nephews. I have five of them and fortunately, I'm on good terms with all of them. Maybe someday they might be curious about what their uncle did for a living and why he never seemed to have a job. (Little did they know I was working from home, sometimes 12 to 16 hours a day.) And what about all the essential pioneers whose names don't appear in any book or if they do, their contributions are understated, or, in the case of people trying to get information from ChatGPT, obliterated? What started as a "chapter or two" on the front end of a collection of articles and talks became a runaway train.

As I was working to recount the various stories I experienced firsthand, I realized that I only knew the parts that I'd been personally involved with so I used the Internet, especially the Wayback Machine (which fortunately was still up then), to fill in the blanks. In the process, I invariably found that the details of the stories of the pre-boom days of the Internet that I experienced were even more fascinating than I realized. Also, very often bits and pieces of a particular narrative were scattered across a dozen or more buried or long-lost articles and had never been put together in a comprehensive form.

In short, I started out with the intention of quickly dashing off a few old war stories. I ended up with the only book I know that goes into granular detail about the Internet's history from 1993 to 1995. Further, I consciously attempted to cover the story broadly, and not just focus on one part or another of the technological, business, financial, or social analysis aspects of it. I hope this book will inspire scholars to take a closer and deeper look at this unique time in history.

Now, back to the narrative.

In 2000, the stock-buying frenzy that was the dot-com boom came to a dramatic end. On April 15 of that year, the NASDAQ, where most of the Internet stocks traded, lost nearly 10% in one day. That's not unheard of for a single stock, but for an index, that's a very very bad day. And it wasn't just a retracement on the way to new highs. Within a month, the index was down another 25%. By the time the bleeding stopped fourteen months later, the NASDAQ was down 78% and $1.75 trillion in market cap (the stock market valuation of a company) went up in smoke.

It would take 15 long years for the NASDAQ to recover.

Giants were battered. Amazon crashed from $113 to $6 per share. Yahoo!, which had reached $475 (adjusted for splits on the way up), bottomed out at $8.11 per share. Cisco, whose hardware ran the Internet, dropped from $79 per share to $9.50. And those were the companies that survived. Hundreds (thousands?) of Internet ventures, public and private, simply disappeared.

Some of the most famous busts included TheGlobe.com which went from a $7 opening price to a peak of $97 on its first day of trading. In less than two years, it was booted off NASDAQ for trading for less than $1 a share. Disney put nearly $800 million into a project called Go.com and wrote it off three years later. Then there was Flooz.com which, shades of crypto, attempted to create its own currency which could be spent at Tower Records, Barnes & Noble, and Restoration Hardware. They raised $35 million, spent $8 million of it on a TV ad campaign featuring Whoopi Goldberg, and then disappeared off the face of the earth in August of 2001.

Pixelon.com promised the world streaming video in 1999 and delivered nothing. The founder, "Michael Fenne", actually David Kim Stanley, was a convicted felon involved in countless criminal scams who was on the run from charges in Virginia. He successfully raised $30 million for his "concept" and threw what is reputed to have been a $16 million party ("iBash 1999") for himself at the MGM Grand in Las Vegas. The party featured Kiss, Tony Bennett, and the Who among others. Not the music of these groups mind you, the actual bands. So much for the due diligence of his investors.

Two of my favorite ventures, because they give a flavor of how separated from reality investors had become, were Flake.com and Seth Godin's plan for a game show network to be distributed via e-mail. Godin got $4 million for that idea, the same amount John Doerr had put into Netscape to get it started. As for Flake.com, it might be an apocryphal tale and I can't seem to find evidence that it actually existed, but the idea was that it was going to be a portal (portals were popular back then) about all things related to breakfast cereal. There were a lot of deals like this floating around in those days.

And finally, there was the magnificent Pets.com which spent $1.2 million in 2000 on a Super Bowl ad featuring a sock puppet, raised $85 million from its IPO the following month, and then went on to lose $147 million in the following nine months. It closed its doors in November 2000. Pets.com was in good company that year. One-third of the ads for Super Bowl XXXIV were dotcomers, fourteen in total, including such "greatest hits" as epidemic.com, e-stamp.com, and onmoney.com.

The Super Bowl wasn't the only place dot-com companies ran ads during the boom. Here's a short list of companies that bought TV spots: Amazon.com, Yahoo.com, Excite.com, Lycos.com, Netscape.com, Snowball.com, eToys.com, Webvan.com, Rent.net, CDNow.com, Vehix.com, and something called gotajob.com which appears to have disappeared without leaving an Internet trace.

Where was I when all this carnage took place? Bettina and I had left San Francisco in late 1998 to move back East.

Things in San Francisco in general and in the Internet industry in particular seemed to me to have gone absolutely crazy. I was, and try to be, simple-minded. I believe that businesses are supposed to make an attempt to sell things at a profit. I believe that businesses should be in business for several years and have a track record of operating profitably before underwriters should offer their shares on the stock market to retail investors (i.e. the public.) Clearly, I was out of step with the "New Internet Economy".

Before I left, I tried to explain my point of view to friends in the Internet world and they all had elaborate explanations for why the valuations made sense which mostly involved the projected growth of Internet use. I had no trouble imagining massive growth in Internet use. What I did have trouble with was the proliferation of truly lunatic ventures and the fact that anyone with even the flimsiest of ideas could find $5 to $10 million to roll the dice on it.

What on earth was going on? I pieced it together years later by following post-crash court cases.

Interest rates were low. Capital gain taxes were low and going lower. There was a lot of investable money sitting around waiting to play. The Internet came along and it was a hot story (with good reason). Netscape's IPO showed the way. It was in business for less than two years, with no profits, and selling a product that may or may not even succeed financially.

From that point on it was off to the races. Not knowing who was going to win, but seeing that a win could be 10x, 20x, 100x, or more, venture capitalists and other investors threw money at everyone with an Internet idea who could fog a mirror. A huge cohort of young people was given large amounts of cash with instructions to "spend it." As I mentioned in a previous chapter, the money went to PR and ads, both online and traditional (like TV), all in an attempt to raise the company's profile to raise more money to raise the company's profile to raise more money.

The brass ring was to get "big" enough - in public profile, not profits or even sales - to attract the interest of the true vultures of the game: the investment banks with their epically corrupt buy-side analysts. "BS.com looks great!" With the public thus primed, the banks would underwrite the stock's launch. Then things got even more "interesting".

Court cases revealed, that some investment banks had made the following deal with their big institutional clients:

"Hey, buddy. We'll sell you big blocks of this stuff at the initial offering price. You hold it to create a false impression of demand. Then when the retail investors (i.e. the public) come pouring in in a frenzy of desperation to get their piece of the action and bid the price to the moon, you can start selling it to them. Just kick back 25% of what you make to us and we'll let you in on every deal we have."

Manipulative stock market operations like this are as old as the hills. There was a previous "boom" in 1969 around semiconductor stocks which helped build out Silicon Valley in the first place. There was a speculative frenzy in radio stocks in the 1920s, in railroad stocks in the 1840s, and before that there was John Law's 1717-1720 scheme to sell shares in Louisiana when New Orleans was a malarial swamp with less than 10,000 people, most of them soldiers with a liberal dosing of convicts, prostitutes, and indigents that France shipped there to remove from the country.

All these propositions eventually panned out in the very long run but the price the public paid to play in the early days was ruinous, so ruinous that in the case of John Law's deal in France, the country, completely traumatized, banned paper money until after the French Revolution, nearly 80 years later.

Did anyone get rich during the dot-com boom? Absolutely.

At the top of the pile were the investment banks and their institutional investor partners in crime. Commercial real estate thrived in San Francisco and elsewhere. All those newly minted companies needed places to play. PR firms thrived. Their services were essential to pump the publicity machine. Traditional media like TV booked millions of dollars in dotcom ads mostly from companies that were too dumb to negotiate rates. Party producers, entertainers, restaurants, luxury car dealers, and companies that sold computer gear and office furnishings, profited too.

The list was a long one and included corporate web consultants who charged hourly fees like attorneys to businesses who were in a hurry to get online fast. Kids fresh out of school working for the big firms billed $300 per hour or more for their time which usually involved having their clients spend way more money than they needed for the services they received from various vendors related to the consulting firm.

Immediately after the crash, at his annual investors meeting in Omaha, Warren Buffet called the dot-com boom "one of the greatest wealth transfers in history". He didn't elaborate, probably in deference to his investment banking pals, but what I described above is what he was talking about. He knew, but as he often does, he left it up to the public to read his tea leaves and figure it out for themselves. They don't call him "The Oracle of Omaha" for nothing.

Where was I? Back east renovating a house and putting in a garden with Bettina, pie-eyed at the absurdly low price of gold, and waiting for people to come back to their senses. I was also amazed at what seemed to be the giveaway prices of real estate in the Hudson Valley in 1998.

When Bettina and I left San Francisco in 1998, $300,000 would buy a bungalow in an East Bay neighborhood where the main industry was slinging crack. We could have bought three or four nice houses in the village where we landed for that amount. I half-jokingly said I wanted to buy the whole street, but there was one problem. No one wanted to sell. Twenty-six years later every one of our three neighbors across the street are still in their homes, and of the five homes to the left of us only one has changed hands.

I waited for the crash, and like all crashes, it started slowly and then happened all at once. Our web hosting, design, and management business didn't lose a single client. Why? All our folks had a legitimate need for having a website and we showed them how to operate online in such a way they made actual money. One of our greatest hits from this time was a client who cornered the market of selling insurance to youth sports leagues (a bigger business than you might imagine) because he was the first online.

In April of 2000, before anyone realized that the "retracement" of Internet stock prices was the first leg down of a major collapse, a friend from the direct marketing industry who'd taken a job at Wharton helping run their program for small business, invited me to come down and give a talk. I remember at the last minute creating a slide that showed the price action of radio shares in the 1920s. RCA scooted from $5.28 per share in 1921 to $568 per share if you account for all the stock splits. This is when $500 was a 10% down payment on a modest house in many parts of the country. If you waited until May of 1932, you could have bought all the shares you wanted for $2.65 each.

When the inevitable question of what Internet stock to invest in came up - I was there to talk about direct marketing online, not give investment tips - I put up the slide and said: "This is what happened to radio stocks in the 1920s. I think the same thing is going to happen in Internet stocks. And you'll be able to buy all the Internet stocks you want for pennies on the dollar." I thought I was overstating the case a bit for dramatic effect, but I was right on the money.

Did I scoop up shares at the bottom? No, I did not. The Enron and Worldcom frauds (remember those?) did not improve my confidence in Wall Street. Two of the biggest companies in the U.S. had been cooking their books and running scams that would have made even the most aggressive Internet stock promoter blush.

Then 9/11 happened.

Between the financial impact of the dot-com crash and the psychological impact of 9/11 which happened in 2001, the economy was on its knees. Live events that required travel were especially hard hit. Ad:tech, the online advertising industry's top conference, nearly went bankrupt. Even months after 9/11 when flights resumed, if you flew then, you and a few fellow passengers often had the entire plane to yourself. Need three seats to take a nap? It was never a problem. Hotels were desperate and willing to deal to book conferences.

San Francisco was like a place that had been hit by a neutron bomb. The buildings were still standing, but all the people were gone. From 1995 to 2000, renting an apartment in San Francisco meant waiting on long sidewalk lines for showings, submitting your credit application with a check, and crossing your fingers. As late as 2003 when I came back for a visit, "For Rent" signs were everywhere again like they were when I moved to the city in 1990.

I decided this was the perfect time to revive my thwarted 1990s dream of teaching my direct marketing on the Internet model to the world.

Chapter Nineteen

There's a System to it

"It must be considered that there is nothing more difficult to carry out, nor doubtful of success, nor more dangerous to handle, than to initiate a new order of things."

— Niccolo Machiavelli, *The Prince*

Now that the Web won, what next?

Here's some of what I did after 1995. As it always seems to work out, one thing led to another.

In 1997, when I was doing my level best to throw sand in the gears of Willie Brown's political machine (see Chapter 17), I was casting around for ways to drive traffic to my website documenting the 49er Stadium Bond Referendum election fraud. Useful e-mail lists of any kind were few and far between in those days and I reached out to Phil Kreutzer, a colleague from the direct marketing world who was one of the very few who knew anything at all about the Internet.

Phil said he didn't have any lists, but gave me the name and contact info of someone who might, and thus I met the wonderful and razor-sharp Jonathan Mizel. Originally from Northern California, he had learned direct marketing by very successfully selling insurance to and through associations. He had already been experimenting with marketing online through CompuServe. In the pre-Web '90s he saw a demo in downtown San Francisco of the "info-bot," probably put on by Sun Microsystems, and he was instantly smitten.

An info-bot is a program that sends people a canned automatic e-mail reply when they write to a specific address. As part of the demonstration, an e-mail was sent to Dee Dee Myers, Bill Clinton's press secretary, and she "replied" right away by e-mail. The audience was astonished. "You know Dee Dee Myers so well that she responds to your e-mails right away?" someone in the audience asked. The presenter replied, "No folks, it's an info-bot and you can have one too."

Like cavemen seeing someone make fire for the first time, the audience gasped. Info-bots are so common now that this may seem ridiculous, but at the time, it was the most space-age online marketing tool imaginable. It's still pretty amazing, especially if you use it intelligently which even in 2024 hardly anyone does.

Jonathan then stumbled into a well-paying job traveling the country teaching the Internet and the then-primitive marketing tools available on it to good-sized audiences all over the country. In the process, he built a large network of people who were very early to put the Internet to business use. When the Internet took off for real, he was in a perfect position to sell banner ads and broker other promotional opportunities to websites that needed traffic. When the dot-com gravy train ended and banner ads crashed from $50 CPM to 50 cents (and lower) CPM,[36] he switched hats and started buying cheap traffic and sending the traffic to well-crafted online offers of various sorts.

While doing this he was also publishing *The Online Marketing Letter*, a print newsletter he started way back in 1993. Then he started hosting *The Web Marketing Power Summit* in Boulder, Colorado where he lived at the time. The name might sound a little hype-y, but it was a carefully curated audience of serious online marketing players with

---

36   CPM means cost per thousand. A $50 CPM means you're paying $50 to put your ad in front of 1,000 viewers. A 50 cents CPM means you're paying 50 cents to put your ad in front of 1,000 people.

very high-level content. When I went to my first one in 1999, the very first couple I met when I walked in the door had just sold their online dating service to Match.com for $16 million dollars. There were a lot of people at his event like them.

The audience was made up of people who had figured out and embodied the concept of "lean-start up" a good ten years before it became a buzzword in Silicon Valley. They had zero interest in going public or taking on investors. They made money the old-fashioned way: They sold products to consumers for a profit and did it as intelligently and efficiently as they could without needing massive infusions of cash. These were the people I'd been longing to meet, but never did, in San Francisco.

Two people I met deserve special mention: Declan Dunn, a very early multimedia-developer-turned-web-pioneer, who wrote *Insider's Guide to Affiliate Programs*. Corey Rudl (1970-2005), who was the smartest and most positive person in a room filled with smart and positive people. Even though Corey was the "kid" in the group, we all looked up to him. His early death still hurts.

Jonathan gave me a chance to speak on the last day and I was all nerves. "What am I going to say to these people who already know and have done so much?" I decided to regale the audience with tales of the origins of Netscape, the company whose pioneering efforts are what ultimately put them in the room and made their new and prosperous lives possible. Everyone in that audience knew who Marc Andreessen was and they were fascinated to learn some inside baseball about him and the twists and turns it took to get the browser out of the cornfields of Illinois and onto millions of people's desktops.

The talk was very well received and I had about 90 brand-new friends, which I think was the cap Jonathan had put on the meeting size. One of the hot topics was

GoTo.com, the world's first pay-per-click search engine. Under this new system, advertisers would only pay when someone clicked on their ad. Bill Gross first presented the idea at a TED conference in February of 1998. People hated it. "It's going to ruin search engines."

In actuality, this is the model that made Google possible. In fact, if you take away the financial underpinnings of pay-per-click, businesses like Google, Facebook, and countless other websites that billions use every day, disappear overnight. Originally, the minimum bid to show your ad on a relevant search result page was one cent. Few realized it at the time, but this was the gift price of the century. Here's what a click can cost in certain categories today. Personal injury: $160 per click and up. Insurance: $50 and up. Loans: $40 and up. When GoTo.com raised their minimum bid to five cents per keyword, some folks lost their minds.

One speaker, Jon Keel, had made a study of GoTo.com's pay-per-click system and developed a methodology business owners could follow for tracking which keywords paid and which ones didn't and at what price. Based on things he learned from Patrick Anderson of ADNet, he discovered that you could almost always find variations of keyword searches that others in your market had overlooked and snap them up cheap. Chris Anderson (no relation), then editor-in-chief of *Wired*, was given credit for coining the term "long tail" in relation to marketing in 2004, but Jon Keel was already using the phrase and concept in 2000 and may have been using it even earlier.[N]

The more I thought about the implications of pay-per-click search marketing, the more excited I became. Pay-per-click put highly targeted and relevant ads in front of people - and it did so without violating their privacy. For example, if I search for "fun things to do in New York City" and up comes an ad for a New York City hotel, it's information that's immediately useful to me and it delivered it without "tracking" me or sending data about me to a third party without my knowledge or explicit permission, something that companies like Google, Facebook, and others now do routinely.

Also, this system would allow people who buy advertising to track the financial performance of every single keyword purchased, again without violating anybody's privacy. If an ad worked and paid for itself at a certain price, you knew to keep running it. If it didn't pay, you stopped running it. This was direct marketing at its cleanest. In 1994, when I pointed out that the click was the ultimate commodity in Internet advertising, and therefore Internet publishing, I hadn't envisioned being able to track ads with this level of precision, but when I realized that it was possible, I was in love.

Jonathan and I collaborated on a program we called *The Anatomy of an Internet Marketing Roll Out* which we released in 2000. Our goal was to give small business people a step-by-step way to first test online marketing ideas and then, if they proved promising, to incrementally increase the ad spend and increase the media used in a systematic way. Too many would-be Internet entrepreneurs were thrashing around without a plan and people "betting the farm" on a single idea was an epidemic, as it is in every industry. Jonathan and I both came from environments where media was brutally expensive (direct mail) and we had this step-by-step, test-as-you-go ethic ingrained in us. The Internet at the time had none of it.

When Jonathan told me he was no longer going to host his Power Summit (it was fun, but too much of a hassle to organize after he moved from Colorado to Maui), I thought about expanding our roll out concept into a full two-day training. I enrolled Jon Keel to teach the pay-per-click part of the course and I focused on strategy, direct marketing principles, and ad copy. I trained some other people to fill in slots because at the time I was dealing with a new and even more serious back injury than the one I'd had in 1988. I just couldn't stand on my feet for long and I needed other speakers to spell me.

We did a trial event in the Washington DC area in the summer of 2001 and it went very well. Then 9/11 happened and as soon as I got over the shock,[37] I started planning the full-blown seminar. I called it The System Seminar, inspired by my friend Vladimir Vasiliev who had a martial arts school in Toronto where he taught a Russian martial art called Systema.

I offered the first System Seminar in the spring of 2002 and I picked Cincinnati as the place to hold it. I did so for several reasons: 1) it's midway between NYC and Chicago, the two big direct marketing cities, 2) it's the home of Procter & Gamble, the world's biggest buyer of advertising, 3) it was a cheap place to hold a meeting, especially in the spring of 2002, 4) I wanted people to focus on the content and not be distracted as they might be in Las Vegas or Orlando, two popular conference sites, and 5) Jon Keel lived in Cincinnati and he and his business partner were handling the hotel arrangements.

Here's what I taught:

The Eight Principles of The System

1. Use A/B split testing to test business ideas, headlines, offers, pricing, and even things like book titles

2. Create a compelling opt-in page and offer to inspire visitors to join your subscriber list

---

37   When I worked for Bankers Trust, which was connected by a pedestrian walkway to the South Tower, I often ate my lunch in a wonderful cafeteria in the South Tower that had a stunning view of the Hudson River. I had good friends who lived in the neighborhood and we spent many a summer weekend afternoon at events on Time Beach, which is what we called the sand landfill on the Hudson River before the American Express Tower was built on it in 1986.

3. Follow up relentlessly with e-mail

4. Develop specific sales pages. Long copy works. Learn how to write it.

5. Up-sell, cross-sell, down-sell

6. Track and test key elements and constantly seek ways to improve your results

7. It's all about traffic and conversion

8. It's a lifetime relationship, not a hit-and-run

1. Use A/B split testing to test business ideas, headlines, offers, pricing, and even things like book titles

Tim Ferris chose the title of his breakout bestseller *The Four Hour Workweek* which came out in 2007 using this method. He simply put pay-per-click ads on relevant search results, testing to see which headline, in his case book title, yielded the most results.

2. Create a compelling opt-in page and offer to inspire visitors to join your subscriber list

As late as 2002, most websites were doing nothing to build e-mail subscriber lists, and those that did were using hackneyed offers like "Join our list". That might have worked in the very early days of the Web, but it had worn out as an effective offer by 1996-97.

3. Follow up relentlessly with e-mail. (In 1995, I started working on an auto-responder program called the Ultimate Mailer.)

It was inspired by the brilliant John McLaughlin who, in the early 1990s, single-handedly and in his spare time as a Sun systems engineer, managed an e-mail-based publication with a massive-for-its-time segmented e-mail list for Sun

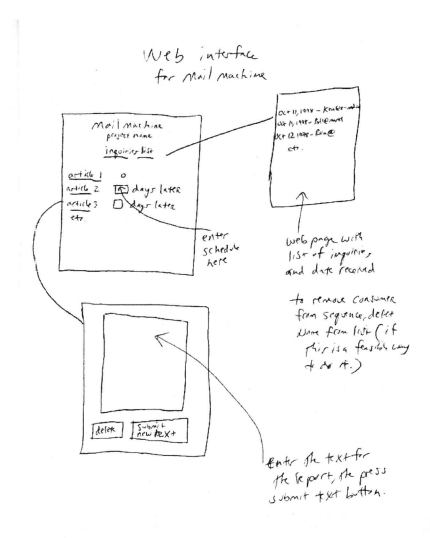

The original sketch I sent to my programmer, Anders Brownworth, for the Ultimate Mailer. Software development is hard, at least for me. This one never took flight, but I did create an early Internet video website authoring management platform that in its time did very well. (Circa 1995)

Microsystems. (My interview with him appears in the back of the book, Appendix III.)

My idea was that when someone signed up for your list, you'd not only send them a welcome message, but you could also send them an endless stream of follow-ups which I envisioned as relevant, interesting, and sales-provoking messages programmed one time that could address an unlimited number of people regardless of when they subscribed. In 1997, a company called AWeber took me out of the misery of trying to develop the thing myself and I became a happy customer and still am all these years later. As common "bread and butter" as this might sound now, in 2002, virtually no one was doing this and a good percentage of the people we taught didn't know it was even technically possible.

4. Develop specific sales pages. Long copy works. Learn how to write it.

In 2002, most people were sending prospects to their home page and letting them figure it out. We taught people to create pages designed to sell specific offers. This is another thing that seems obvious and is as common as water now (video sales letters fall into this category), but almost no one was doing it then. We also introduced Internet marketers to the idea of taking the time to tell your *whole* story. "But no one is going to read long copy on the Internet," many told us. Well, billions and billions and billions of dollars in sales generated by long copy say, yes they will.

5. Up-sell, cross-sell, down-sell

The absolute best time to sell someone something else is when they are already in the process of buying from you. It will never get easier. The six words "Would you like fries with that?" have probably sold billions of dollars in french fries and every business, if they look, can do something similar. Even today, far less than 1 out of 10 try. In those days, in 2002, less than 1 out of 10,0000 people selling things online gave any thought to this, and it may have been even far fewer than that.

6. Track and test key elements and constantly seek ways to improve your results

In 199_, I wrote a book called *Testing: The Key to Direct Marketing Profits* that included a detailed description of A/B split testing and I gave a copy to all the students of my early Internet seminars. In 2012, *Wired* published an article that claimed Google discovered this technique in 2000 and with it transformed the way websites work. All I can say to that is, "Give me a break". (I explain how you can get a copy of this book for far less than its current $250 Amazon list price on the last page of this book.)

Test everything. Test one landing page design against another. Test headlines against each other. Test different prices and terms. Test product names. Always keep testing. So-called small improvements, say from a 1% order rate to a 1.5% order rate, mean a 50% boost which could be the difference between a winning promotion and business and one that fails. If you relentlessly keep seeking small improvements, in time it can add up to massive ones.

7. It's all about traffic and conversion

The great challenge of business is focus. It's frighteningly easy to spend the majority of your day doing things that do not lead to sales and profits. Marketing on the Internet and in any direct response medium is a simple business. You need more traffic and you need to convert the traffic you get better. Thinking about anything else is a waste of time. (One early student of The System who had nothing but a website with a few pages on it told me that adhering to this simple principle became the foundation of his 8-figure-a-year privately owned business.)

8. It's a lifetime relationship, not a hit-and-run

Old-school mail-order businesses were happy if they broke even, or even lost a little money, on the first sale to a customer. Why? The arithmetic of their costs in acquiring a customer forced them to recognize the real business they were in. They knew they were not in the one-time sales business. They were in the customer acquisition business. The massive profits for ethical businesses come from the back end, the stream of sales that come from satisfied, well-cared-for customers. A shockingly large number of businesses still don't know this to this day.

When you put all these things together, it was called The System, thus The System Seminar, and if you're approaching your marketing this way there's a high likelihood you learned it from us, a student of ours, a student of a student of ours, or a student of a student of a student of ours.

Speaking of students, one of the things I'm proudest of are the people we taught direct response-oriented online marketing to who went on to become genuine leaders in the field.

Examples:

Perry Marshall, who we taught pay-per-click advertising to. I also personally introduced him to Richard Koch and his profoundly useful book, *The 80/20 Principle*. Perry wrote the *Ultimate Guide to Google Ads*, the first book on the subject now in its sixth edition.

Kim Dushinski, a book publicist who we taught Internet marketing to from the ground up, released *The Mobile Marketing Handbook* in 2009, just two years after the introduction of the iPhone. It was the first-ever guide to mobile marketing.

Ben Hunt, a veteran web designer since 1994, took our seminar in 2008 when we brought it to London, learned our approach to marketing on the Web, and to his credit, completely rethought everything he "knew" about web design. The result was his book *Convert! Designing Web Sites to Increase Traffic and Conversion* which came out in 2011 and it's still the only book on how to use direct response principles to design web pages and websites.

Lloyd Irvin was a Brazilian jujitsu, judo, and Russian Sambo champion in the 1990s and operates a 10,000-square-foot training facility in Temple Hills, Maryland. He leveraged the foundation he got from us in 2003 into several successful businesses that he owns outright and shares in another seventy, which he acquired by trading his Internet marketing prowess for equity. He is also a superb educator.

In 2002, Mike Stewart, a System grad, and I introduced the first push-button audio streaming on the Internet. Previous to that, users had to have an audio player and download whatever they wanted to hear individually before they could listen to it. Mike realized that if you simply put audio alone into a Flash player instead of video and audio, you could put a button on a web page that users could push and get instant audio streaming. A simple idea, but revolutionary for its time. If you have ever pushed a button on a web page and gotten audio, it started with us.

I think the reasons so many of the attendees of our seminars were so successful, besides their own intelligence and drive, were twofold.

First, when I started the System Seminar I had nine years of online marketing experience under my belt and, at that time, I was among the most practically informed people on the subject on the planet. And if I didn't know something, I was one, maybe two at the most, phone calls away from someone who did.

Second, I taught the seminars in the spirit of 1993-1995, when the point was to use education to bring new congenial people on board to join the party. Some of my instructors couldn't grasp this. All they could see when they got on stage were personal dollar signs.

It was both annoying and amusing to see how quickly some of them changed from educators to wannabe cult leaders as soon as they got a little prominence by virtue of being on my faculty. It became necessary to show many of them the door. (I'm toying with putting a list of the worst offenders on the website that accompanies this book, using nicknames to protect the guilty.)

In 2005, after I saw my first demonstration of the video embed code, I started a website called SystemVideoBlog.com where I declared, among other things, that with this development, "video is the new paper." Colleagues thought I was nuts. "No one is going to watch video content made by amateurs." "Nobody is going to watch long videos on their computer."

At that time, there were 5 million videos on YouTube total. Now there are 14 billion and growing. I rest my case. I love YouTube myself (though I have concerns about how Google runs it).

The idea of a corporate Time-Warner-style interactive TV with one entity controlling all the content was always a non-starter for me. In contrast, having access to billions of videos made by people from all walks of life who have good ideas, specialized knowledge, and unique points of view is pretty close to my idea of media heaven.

I'm a child of the three-network television generation.

When I grew up we were subject to whatever ABC, CBS, NBC, and the forces that controlled them, wanted us to think, know, and believe. From an early age, I didn't like this system. As I grew older, I came to believe that unless humanity took control over that box in everyone's living rooms and what it meant, we would forever be living at a fraction of our potential as human beings.

Very shortly after I was first exposed to the online world, I saw a real chance in it for a way out, and could not rest easy until it had achieved its potential. Now I rest easier.

Use your freedom.

Make something and share it.

Whether it's for ten people or ten million, it doesn't matter. And directly support the people who are making things for you. The fewer middlemen we have between creators and audiences, the better. We need centrally controlled information systems like a fish needs a bicycle.

To paraphrase the dancer Isadora Duncan:

You were born wild. Don't let them tame you.

Epilogue
What if?

"How many Microsoft engineers does it take to change a light bulb? None, they just declare darkness the standard."

> – A widely circulated joke from the 1980s among people who cared about the quality of the products of the personal computer industry

It was all inevitable, right?

The Internet as we know it today just sort of happened and the details of how it came to be are inconsequential.

That's one view of things and frankly not a very intelligent one. If you're not aware of how things come to be, you'll be forever handicapped in your own attempt to make things come to be, thus one of my main purposes in telling this story.

That we have the Internet as we know it today is the result of two different chains of events, one technical and one financial. A break in any of the essential links in either one of these chains could have easily derailed the train and if it had, we would almost certainly be looking at a very different online landscape, one of which I will describe later in this section of the book.

The standard narrative is that the government created the Internet infrastructure and after twenty years made it available for personal and commercial use and everything else is random detail.

Missing from this narrative is the fact that on January 1, 1989, when the Internet was declared "open for business" essentially nothing happened. Nothing much of consequence happened on the business front in 1990, 1991, 1992, and 1993 either, with the one exception being BBSs, with CompuServe leading the way in 1989, giving their users access to Internet e-mail. Other than that, it was "crickets". In short, simply declaring the Internet "open for business" by itself did next to nothing.

Netscape's eye-opening 1995 IPO performance changed things from the point of view of the levers of finance. That said, the foundation for the stock's unexpected performance, when its price rocketed from $28 to $74.75 on its first day of trading, was laid in the previous handful of years with 1994 being the one that made the all-important *commercial* difference.

On the technical side...

1. Tim Berners-Lee and Robert Cailliau wrote the code for the World Wide Web and their employer CERN made it available to all in 1991, after insistence by Cailliau that CERN formalize its public domain status.

> *"It was important to put the basic library, the documentation, to make it available to everyone without any strings attached so that it could explode. Americans always worry about lawyers. They cannot understand that anything produced by CERN is just out there, you can use it. As long as they mentioned it came from CERN, we're happy. But they wanted written signed statements from our administration that they won't sue them and so I needed some statements like that in order to get some of the people from some of the other institutes to collaborate with us."*

- Recorded dinner conversation:
Robert Cailliau interviewed by Charles Petrie
Editor-in-chief of IEEE Internet Computing
November, 1997

By the end of the year, a few hundred people were using it and the world saw its first few dozen websites. Everything fundamental about the Web - the HTML to create web pages, the URL format to create web addresses, and HTTP the protocol that makes it all work - was all figured out by the team at CERN.

There was no imperative for Berners-Lee and Cailliau to do the work they did. Nor was there any imperative for CERN to formally make the code available to the world. Had either of these two things not happened, there would be no Web at all and thus no Internet as we know it today.

2. Twenty-two-year-old Marc Andreesen stumbled on the then highly obscure World Wide Web while on a work-study job at the University of Illinois at Champaign-Urbana. His pay on that job was $6 and change an hour. In February 1993, less than a year before his graduation, he and Eric Bina decided to try their hands at putting an easy-to-use graphical interface over the clunky text-only code that was the Web then. They stuck with it, finished it, and most important of all promoted and supported it (it was shareware, free to the public). By the time Andreessen graduated in December of 1993, their creation, the Mosaic browser, had over 1 million downloads.

I'm going to be a little repetitive with my language throughout this section. It's hard to come up with different ways to say the same thing to make the same point, so Homer-like (that would be the Homer of the Odyssey and the Iliad, not Homer Simpson) I'm going to use formulaic repetition.

There was no imperative for Marc Andreessen and his colleagues to do the work they did. Nor was there any imperative for them to make it available to the world and, most importantly, devote hours and hours to supporting users. Had either of these two things not happened, there would be no Web and thus no Internet as we know it today.

3. When Andreessen graduated and arrived in Silicon Valley, the world's capital of technical innovation, a red carpet was rolled out for him based on what he had accomplished with the Mosaic browser. He was written up in the computer industry trade journals. *Wired* wrote a feature on him.

Actually no. None of these things happened. Instead, he was offered and accepted an interesting job commensurate with what any newly-minted computer science graduate could expect, and he felt he was at a dead end in his early career.

What happened next was that Jim Clark, the founder of Silicon Graphics, left the company he started in frustration over its internal politics and its decelerating interest in innovation and was looking around for new opportunities. In his quest, he reached out and talked to many people for ideas. One of the people he talked to, recommended to him by a colleague, was Marc Andreessen.

After considerable conversation with Andreessen, and after a lengthy process of elimination, the two came to the unexpected but ultimately obvious conclusion that the way forward was not interactive TV or doing a deal with Nintendo, which is where the two started. It was making and marketing software for the Web, at a time when that was the most implausible business imaginable. They started Mosaic Communications (later Netscape) on April 4, 1994, with the goal to make web browsers and servers a commercial product.

There is nothing written in stone that said it was inevitable for Clark and Andreessen to meet, much less engage in extended conversations about business ideas that led to the creation of a company together. Clark could have gone in another direction entirely or he could have taken that spring off for an extended sailing voyage at sea, something he's done often.

Had Clark and Andreessen not met and had a meeting of the minds that led to the formation of Mosaic, there would be no web and thus no Internet as we know it today. No other player possessed anything even remotely like the horsepower the Netscape team had to successfully propel a web browser/server software company into the mainstream.

This brings us to the end of the critical chain of events on the technical side. If you do the simple thought experiment of removing any of these easily removed links in the chain, you can see there would have been no credible commercial players in the web browser/server space and no spectacular 1995 IPO. Thus, the Web, which had been an obscure protocol, could easily have sunk into oblivion as did so many other attempts to make the Internet easy to use and ready for prime time.

Now on to the purely business side of things...

Business is based on making products and services available to the marketplace for a profit. No profit - or hope for profit - and there are no funds for investment and re-investment.

Building out the Internet infrastructure so that high-speed access became the norm required many, many billions of upfront dollars. Until Netscape's IPO kicked off a Wall Street investment firestorm, the amount of investment money available for Internet companies was minuscule, nowhere near what was needed to build out what we have today.

While it's true much of that money was wasted in fraudulent underwriting operations (shout out to Goldman Sachs, Merrill Lynch, Morgan Stanley, Credit Suisse/First Boston, Lehman Brothers, and Deutsche Bank), investment for all things Internet flowed like water from August of 1995 through March of 2000, long enough to give the Internet the financial and publicity kick start it needed to reach critical mass. Without the Netscape IPO happening the way it did and the manic gold rush that followed *there would be no Web and thus no Internet as we know it today.*

That's what happened from 1995 to 2000, but what happened in 1994 and in the months leading up to the IPO that prepared the public for the event that started the gold rush in the first place?

It wasn't Netscape's profits in 1994 and 1995. There were none. In fact, it never made a profit as a standalone company and eventually sold itself to AOL in 1998. So it wasn't Netscape's numbers that excited Wall Street. It wasn't computer ownership either. In May 1994, the Times Mirror Center (renamed the Pew Charitable Trust) poll estimated that only 12% of American households had a computer with a modem and three months after the Netscape IPO a follow-up study by Pew found only 3% of American households could access the Web.

Some may point to the founding of Amazon (July 4, 1994) or the founding of Yahoo! (January 18, 1994) as being significant. While these companies excited a few intrepid venture capitalists about the long-term money-making potential of the Internet, in 1994 and 1995 these companies had no profits either.

While these were interesting ventures for the time, they were nothing special. In 1994, Amazon only sold books and didn't get around to offering products other than books until 1998. John Doerr of Kleiner Perkins, who was an early backer of Netscape and other Internet icons, didn't start paying attention to Amazon until the spring of 1996. Amazon didn't show a profit until the fourth quarter of 2001, and it was a modest one at that, just $5 million. As for Yahoo!, it didn't have any revenue at all in 1994 and only started selling banner ads in August 1995. Yahoo! didn't show a profit until the fourth quarter of 1997.

If you're thinking about Internet giants like Google and Facebook being a factor, they didn't even exist until 1998 and 2004 respectively, and didn't start monetizing their traffic – selling clicks - until 2002 for Google and 2006 for Facebook.

What happened in 1994 that changed the game and investors' perspectives so dramatically?

First, let's be clear about what it wasn't.

It wasn't the market penetration of the Web. Web use was puny compared to other online options at the time. It wasn't hype from the news media. *Wired* didn't even mention Marc Andreessen and Netscape in its pages until October of 1994, and failed to mention the Internet at all in its first issue. It wasn't the plethora of profitable Internet companies. There were none. They were all bootstrap dreamers which included Amazon and Yahoo! at the time. If it wasn't those things, what was it?

I think the case can be made that it was the lowly banner ad, which ultimately is based on the selling of clicks.

Other people and companies had taken a halfhearted stab at selling advertising on the Web,[38] but the first company to do it in a serious, industrial-strength way was Hotwired.com and its sales team led by Rick Boyce. (You met him in Chapter Eleven).

They were the first company in history to stick a flag in the ground and say: "We're going to be a serious Internet-based publishing company, with real offices, executives, employees with salaries, etc., and we're going to pay for it the same way every other publishing entity on earth does it, by selling advertising." This may sound like a small and obvious step, but at the time it was at the level of Neil

---

38   Global Net Navigator (GNN), funded by tech writer-turned-publisher Tim O'Reilly, sold the first clickable web link ad in 1993 to a Silicon Valley law firm.

Armstrong taking one small step and one giant leap for the financial viability of the Web.

To understand the significance of this, all web history can accurately be divided into before Hotwired.com and after Hotwired.com. Before Rick Boyce and his sales team, commercial publishing on the Internet was nothing but a dream. After Rick's team sold out all the ad spaces in the first issue of Hotwired.com, it became "Here's how we're going to pay for this thing." It was a defining moment.

Ad revenue is the financial fuel that keeps the Internet running. Consider that Google and Facebook are, all the PR and uncritical news media reporting aside, simple ad-supported publishing businesses. They provide content (other people's content). It attracts an audience and they sell access to that audience to advertisers. That's where their money to operate comes from. Without ad impressions and clicks, there'd be no Google or Facebook. Or Yahoo!, or Instagram. Or TikTok. Or YouTube. Or millions of lesser-trafficked websites. Poof. They'd all go up in smoke overnight.

Even Amazon gets a substantial slice of its operating income pie from selling ads on its site, $41.95 billion a year worth as of a March 2024 forecast. And unlike its impressive, but expensive infrastructure of warehouses, packing and shipping facilities, and staff and the low margins it makes from selling products (the business that most people think Amazon is in), most revenue made from selling ads goes straight to their bottom line. It goes a long way to helping keep the lights on.[39]

Search engines made the Web usable. Without them, we're back to an analogy I presented earlier in the book:

---

39   While we're on the subject of Amazon secrets, as of the first quarter of 2024, 61.5% of the company's operating income (the money that pays the bills) comes not from selling products, but from selling web services to other companies large and small. It was hard to make a profit selling products online in 1994 and it's just as hard today, thirty years later.

the world's largest library but with all the books in a great big and totally random pile. Interesting, but not remotely useful.

The Hotwired.com sales team's blockbuster success selling advertising to deep-pocket advertisers was the shot heard 'round the world or to use an analogy from the Iliad, the face that launched a thousand ships. Once Hotwired.com demonstrated that websites with audiences could get checks from the likes of AT&T, IBM, Volvo, and Saturn, the Web gold rush was on. It was a simple model even the most unimaginative could understand and launch without too much capital: put up a website, hire some writers, and sell ads.

Some reading this might say, if it wasn't Rick Boyce surely someone else would have come along and gotten the ad sales ball rolling on the Internet. Yes, that's probably true, but who and when, and *would it have happened in time?* Even with Hotwired.com's example, it took Yahoo! ten months to get with the program. The then-important search engines Excite and Lycos took seven. Their slow motion is puzzling because ad sales are the only way these services ever made any money.

My question, "Would it have happened in time?", is not an idle one.

Given the negligible presence of the Web in people's homes, (still less than 3% even right after the Netscape IPO), the lack of corporate investment, and the fading interest of the easily distracted news media (which was already clickbait-driven in its DNA even before there were even clicks to sell), it was entirely possible that the Web could have fizzled out before the 1995 Netscape IPO. And then what? No Wall Street investments, no build-out, no critical mass.

If we didn't get the Web, what might we have gotten instead? To answer that, we'll have to engage in some speculative history and play the game of "what if?" For example, what if you're mother never met your father?

That's easy. You would not be here reading this. Another "what if?" that people, including even professional historians, like to play is, "What if the Nazis won World War II?" Hollywood movie heroics aside, this is not such a far-fetched scenario.

In 1939, the physicist Albert Einstein addressed a letter to "F.D. Roosevelt, President of the United States" expressing his concern that the progress that Nazi Germany was making in nuclear physics could lead to the development of viable atomic weapons. Roosevelt could have crumpled up the letter and tossed it or filed it for future reference. He didn't. He started the U.S.'s own massive covert program to develop the nuclear bomb.

The Nazis were working hard on their atomic bomb and they had a head start,[40] but as luck would have it, the U.S. developed its bomb before the Nazis did theirs. Had the Nazis been first, cities like Liverpool and Birmingham ("the city of a thousand trades") might have been the first victims of nuclear warfare, not Hiroshima and Nagasaki. It's not clear how Churchill's stiff upper lip would have held up to that. Without Britain as a safe base, there would not have been a D-Day. The rest of the continent, and much of the eastern hemisphere could have been overrun by overt Nazis and other flavors of fascists.[41]

What if the Web stumbled and fizzled in 1994? Then the online world would have continued on the course it was already on with multiple proprietary services that charged

---

40   A company owned by the father of Klaus Schwab, a protege of Henry Kissinger and the founder of the World Economic Forum, provided the Nazis with the precision tubing for use in their nuclear weapons program.

41   Instead, we have Ursula von der Leyen and the European Union.

by the hour for access: CompuServe, AOL, Prodigy, GEnie, and tens of thousands of quirky smaller ones.

Here's what this would have meant in practice:

Want to post content online or start a discussion board? Submit your application to corporate headquarters.

Want to sell products and services online? Contact the advertising department for rates.

Want to run an online store? Sure. Bring lots of money in fees and be prepared to conform to whatever half-baked system we've cobbled together and deal with whatever dimwits we've hired to manage it.

Want to sell to *everyone* online? Gee, aren't you ambitious? Can't help you there, but you can sell to our users on our proprietary system and once you've run that obstacle course, you're free to contact all the other online services and work it out with them. Good luck.

That would have been our online world. It's obvious that under this kind of regime, many millions of online ventures large and small from Amazon on down would never have come into existence.

But it gets worse.

Before the Internet boom, the personal computer industry which had fueled Silicon Valley and San Francisco during the 1980s, had run out of stream. People who felt they needed personal computers already had them, and there was no big imperative to upgrade to new gear.

Thus the "hope-ium" that surrounded the multimedia industry in the early 1990s. Hardware makers and their software allies believed that somehow multimedia titles on CD-ROMS were going to inspire people to buy faster machines. That wasn't happening.

After launching the Macintosh in 1984, the first point-and-click graphical user interface for personal computers, Steve Jobs was in effect booted out of Apple in 1985, the company he started with Steve Wozniak in 1976. Jobs left the personal computer business and took up developing sophisticated workstations (NeXT) and animated movies (Pixar).

For reasons still unclear to this author, Apple was put in the hands of a guy whose previous job was selling carbonated sugar-water (John Scully formerly of Pepsi) and the company was dying the death of 1,000 cuts. Many were justifiably concerned that the company would not survive. Jobs would not return to take Apple off life support until 1997.[42]

Unless you lived through those times (the early 1990s), it's hard to imagine the extent of the dominance of Bill Gates and Microsoft over what appeared on people's computer screens. If you were one of the relatively rare people who was not using an Apple computer (well under 10% in 1994), you were in Bill Gates's pocket. Microsoft had the monopoly on computer operating systems and all the major software programs that ran on it. In a world without the Internet, what Microsoft allowed on your desktop made up the sum total of your computer experience.

In the early 1990s, Gates had two obsessions, CD-ROM publishing and the launch of Windows 95, with the Windows 95 launch taking up 95% of his attention. It was Microsoft's first attempt to improve its operating system and graphical user interface which was, to sum it up in a word, pathetic. Ten years after the original launch of Windows, Gates, after an 18-month delay, was hoping to wow the world with Windows '95.º

---

42   I was eating dinner in a Japanese restaurant on Fillmore Street in San Francisco with the Internet industry's top PR mastermind Ed Niehaus and his family when he got his first pager message from Jobs: "Call me. Steve."

Gates had little interest in the Internet in 1994, so little that his book *The Road Ahead* on the future of computing published in 1995, barely mentioned it in its first edition. Instead, he made frequent references to the "information superhighway." By the spring of that year, he seemed to get a clue and wrote a long memo to his executive staff which among other things pointed out that "after 10 hours of browsing, I had not seen a single Word.DOC." The Web and the attention it was getting, especially after the Netscape IPO, became an irritation to him because it was stealing the thunder of his launch of Windows '95.

And what a launch it was.

Gates reportedly budgeted $300 million for advertising to launch Windows 95 including a $3 million fee (some say $8 million) to use the Rolling Stones song, "Start Me Up" for its advertising. It was the first time the band had licensed its music to an advertiser. This level of ad spend buys a lot of news media influence.

Microsoft already had effective control over the personal computer press. Publishers and editors were used to getting phone calls from and being personally harangued by Steve Ballmer. Bad press for the company, regardless of how often it stumbled or how bad its products were, was essentially forbidden if you wanted to keep getting Microsoft checks and be free of Ballmer's tirades.

The owner of one of Silicon Valley's top PR firms in that era told me, "Our job is to get innovative start-ups the attention they need so they can stay alive long enough for Microsoft to either buy them or destroy them." It was not at all unusual for the superior products of companies that Microsoft "handled" to be buried, never to be seen again. That was the state of the innovation in personal computing pre-Web.

As Microsoft was planning its Windows '95 launch, it was also preparing to launch Microsoft Network (MSN).

It was, like many Microsoft-originated products, garbage, and the "concept" was to steal market share, and, if true to character, undermine fast-growing AOL along with the other big online services. Unfortunately for Microsoft, AOL stole a march on them, and became too big to crush, leaping from 1 million to reportedly 20 million users from 1994 to 1995.[43]

Gates's plan was to put access to the Microsoft Network right on everyone's desktop so it would be the first thing you'd see when you logged onto your computer. On over 90% of all desktops in the world. No $300 million worth of free-trial CDs needed, the amount AOL had spent to promote itself. Microsoft planned to follow the industry standard which was to charge for online access by the hour. AOL didn't break that model until December 1996 when it went to a flat monthly rate of $19.95 per month.

Let's engage in a little speculative history.

Imagine instead of waiting to launch Microsoft Network in 1995, Microsoft managed to get itself together to have it up and running in 1993 *before* AOL started its growth spurt, *before* someone figured out how to generate meaningful revenue on the Web by selling ads, and *before* Netscape had nearly a year and a half to make itself into a media darling paving the way for the 1995 IPO that changed everything.

AOL and the Web could have been strangled in the crib and thus we'd have Microsoft-level innovation in features and interface, i.e., crap. Access to online services could have very well remained something you paid for by the hour.

---

43   AOL played fast and loose with its subscriber numbers and other key metrics like ad sales so there is no telling what the real numbers were, but no doubt it was growing fast. At one point, under the direction of AOL's Chief Marketing Officer Jan Brandt, AOL was sending so many AOL-free start-up CD-ROMS in the mail that the company was responsible for 50% of the total CDs manufactured in the U.S. Brandt estimated the company spent over $300 million mailing all those free trial disks.

With Microsoft's near-total control over the desktops of the world and its ability to bend the press to give the public its version of reality ("What do you need to mess with the Web for? It's all on MSN and it's easy."), we could easily be living in a world where the Web was just another attempt to make the Internet user-friendly that failed. In that case, the only access to the online world would have been the Microsoft Network because CompuServe, AOL, Prodigy, etc. would have been crushed and the fledgling web would never have gotten out of the starting gate.

If this sounds implausible, look at the impact a company like Facebook has had on the Internet. With none of the advantages enjoyed by Microsoft and by just making things "easy" for smartphone users, Facebook has siphoned off hundreds of millions of Internet users worldwide who think Facebook's online dystopia *is* the Internet.

I see no reason why Microsoft, if they had acted sooner, could not have convinced everyone, excluding a tiny holdout of die-hard techies, that their Microsoft Network was not only all the online experience they needed but also the *only* online experience that was possible. Paid enough money, all the news media, including *Wired*, would have been happy to go along with the charade.

Thus my assertion that the things that happened in 1994 which no one saw coming and which were barely reported at the time, laid the foundation for what made the Web as we know it today possible: ubiquitous, not under centralized control, and open to anyone with the will to publish.

Without these events, we might still be faced with a handful of major online providers, or potentially just one controlled by Microsoft, that charge for access by the hour, offers limited content, and make independent online publishing and commerce, if not impossible, irrelevant.

Four key events made the difference:

- Netscape became the first credible company to go into business to make and sell web browsers and web servers on April 4, 1994.

- Rick Boyce was exposed to the click-through as an advertising metric at the June 11, 1994 meeting at 3220 Sacramento Street in San Francisco.

- *Wired* management decided to create a web-based magazine (Hotwired.com) with ad sales as its financial engine, (the first venture of its kind).

- Rick Boyce and his sales crew closed the sales and thus the website Hotwired.com launched with a full array of corporate advertisers on the day it opened on October 27, 1994.

The Pentagon and the federal government did not make any of this happen. Their contribution, significant as it was, was solely making the canvas available.

Companies like Microsoft and Apple did not make it happen. At the very time these events were taking place both founders Bill Gates and Steve Jobs respectively declared the Internet irrelevant.

Silicon Valley did not make it happen. As late as 1994, the software companies of the time were overwhelmingly disinterested in creating products for the Web. Some venture capital firms made small bets on companies like Netscape and Yahoo!, but these investments were a small fraction of their business at the time.

Wall Street did not make it happen. It completely ignored the Internet until the Netscape IPO on August 9, 1995.

The news media did not make it happen. Any systematic review of the news media's reporting about the

Internet in 1994 will show their stories were barely coherent and poorly informed, and covered the spectrum from wild-eyed claims to dismissal and ridicule, empty of even the most basic research and critical thinking.

The news media, Wall Street, Silicon Valley, and the federal government did not make the Internet as we know it today. Individuals did, countless thousands of generous, enthusiastic individuals who each contributed a piece of the puzzle in pursuit of a simple and unified vision: to make the publishing and consumption of useful information as streamlined, geographically unbounded, and available to as many people as possible. In pursuit of this vision, they launched a revolution which is right up there with the impact of the printing press and universal literacy.

There was once a time, and it was only a few centuries ago, when individual human insight and experience could only be transmitted locally to a few people at a time. Otherwise all permitted "knowledge" was presented top down with the penalty for failing to accept it, or daring to correct or expand on it, being exile, imprisonment, torture, and even death. The distorting, withholding, and repressing of information was wielded like a weapon; knowledge was used to control not elevate human beings.

All this runs counter to genuine human progress which is based on the accumulation and transmission of knowledge. This impulse, whether it expresses itself in immersion in music, numbers, visual images, narratives, language, the natural world, manual skills, or procedures of all kinds is, like the impulse to nurture the young, fundamental to human existence. It's one of the things that makes life worthwhile and improves the chances of survival for us individually, for our communities, and even for our species.

Anything that interferes with this impulse should be circumvented, fiercely resisted, and ideally defeated.

Anything that fosters it should be celebrated, protected, and defended. Our society claims that it does this, but this claim does not hold up to close examination.

These days, and perhaps it's always been thus, the value of teaching and other forms of sharing knowledge is only formally recognized and valued when it takes place in the context of approved so-called educational and news media organizations. I say "so-called" because with almost no exceptions, especially on the institutional end of the spectrum, these organizations more closely resemble bureaucracies that value conformity and the mediocracy it breeds above all else.

In a climate where conformity and its inevitable byproduct of mediocracy rule, the real pursuit of knowledge, which always includes sharing it with others, is disdained and even actively repressed. In its place are "programs" that fit the ideologies of those who control the bureaucracies. By this standard, if these programs result in the amputation of the potential of human beings and their communities so be it.

In its early years, the Internet promised to break this chokehold on humanity which is why a generation of engineers and then entrepreneurs and genuine educators poured incalculable amounts of time, uncompensated and without even the notion of compensation into its creation. Despite having no clear conception of where success would be or exactly what it would look like, the pioneers kept moving forward relentlessly and as a result, we have what we have now.

Today, a significant percentage of the planet has access to the Web. With minimum procedural guidance, people can learn to find and take advantage of high-quality information on literally every subject that human beings concern themselves with. The cost of this access has been reduced to their own time, energy, and ambition and a connection to the Web. The barrier to information has never been lower and it's hard to imagine how it can be improved on. An interesting statistic: Despite the mass media's obsession with force-feeding the public mindless entertainment and sensationalized ideology-laden "news", according to a Pew Research Center study over half the people who use YouTube, now the second largest search engine on the Internet, use it to educate themselves.

On the other side of the equation, while no one knows how many of the nearly 64 million people YouTube calls "content creators" are educators, a search of topics I'm familiar with generally reveals that it's virtually impossible to get to the end of the inventory of quality educational content. And by *quality* I mean at least equal to and often superior to anything available on any TV channel (sorry BBC and PBS) or in any Ivy League classroom (sorry Princeton and Harvard). And you can fine-tune the info you're seeking with exponentially more precision than you can on any college campus.

This brings us to the question, how is the Internet delivering on its potential?

On the grassroots level, as demonstrated by genuine education-minded people who use YouTube as a platform, not to be confused with what YouTube itself does, it's a success beyond all expectations.

The body of published works – text, audio, and video – that individual educators and small teams have already created surpasses the British Library, the Library of Congress, and the course catalogs of every university in America. It's not even close, especially when you look at the broad spectrum of information available and how up-to-date much of it is.

Marveling at this accomplishment, you have to ask why it's not trumpeted daily. I think the answer is simple.

It would give the public too much credit and highlight the shabbiness of the products of our educational and media bureaucracies.[P]

YouTube is a creation (or strictly speaking an acquisition) of Google. Google bought the platform for $1.65 billion in February 2005 because it was incapable of creating anything like it on its own. As a technical achievement, it's impressive for the reasons I've already given. Anyone with a body of knowledge can share it on the platform for free and with a little business smarts, and a lot of business diligence, use it as the front end to a livelihood, sometimes a good or even a great one. Users can access all the content for free if they can accept having their viewing time interrupted by ads. If not, they can access the entire 14 billion, and growing, collection of videos for a monthly fee.

Clearly, the original ethic behind YouTube is a win for humanity.

Where YouTube falls down is the same place its parent company Google falls down. Google was once an excellent search engine. Now it's a better-than-nothing search engine. What made the difference? Eric Schmidt's (CEO: 2001 to 2011, Executive Chairman: 2011-2017) decision to turn the company into a favored-ideology-of-the-day promotion and censorship platform. The results of the change are obvious in the ever-declining quality of Google's search engine results. YouTube has followed Google's dictates by removing countless worthy videos because they voice opinions and share facts that Eric Schmidt and the people he left in place don't want heard.

Normally, the noticeably reduced quality of a company's product would put a company at risk. But not Google and YouTube. Like Microsoft before it, Google and YouTube have attained monopoly status, and at that level, to quote the 19th-century robber baron William Vanderbilt (son of Cornelius), "The public be damned."

Then there's Amazon. On the positive side, Amazon has created a product acquisition/warehousing/delivery logistics system that is second to none. It does not engage in censorship very often. When many public libraries and most U.S. bookstores would not carry books from qualified experts (including veteran physicians and internationally recognized professors from relevant disciplines) who questioned the COVID narrative, Amazon carried them. Amazon has also made it incredibly easy to get book titles fast no matter where you live and it has streamlined the process for self-publishing to an impressive degree.

On the dark side, working conditions in its warehouses are terrible, it plays a particularly nasty game of hardball against union organizing, and it has engaged in vicious legal attacks to stifle competition to a reprehensible degree. It plays the same acquisition game that Microsoft played. Offer a lowball buyout price and if the target company doesn't agree, reduce it to rubble by selling competing products at a loss for as long as it takes to wreck it.

Amazon gets away with these practices for the same reason Google does, but founder Jeff Bezos added an additional wrinkle by buying the failing, but politically powerful *Washington Post*. Politicians who consider calling foul on Amazon's abuses of its monopoly have to wonder if one of their own misdeeds will become a front-page *Washington Post* story in retaliation. No one does political hatchet jobs like the *Washington Post* and Congress knows it.

Finally, there is Facebook. I'm at a loss as to what this company has ever contributed to the Internet. It did a brilliant job of conning tens of millions of Americans, small businesses, and other organizations into thinking that Facebook should be *the* place for their web presence. Nice for grandma so she can see current pictures of the grandkids, maybe, but disastrous for small businesses, non-profits, and community groups who follow this path.

Why on earth would any thinking person use Facebook as their sole web presence? If you are one of these people, please, and without delay, spend less than $10 and get your own domain name, learn how to use one of the dead simple web authoring programs, and invite your friends and customers to give you their e-mail addresses. Being dependent on this creepy platform is one of the worst things you can do as an organization.

Why creepy? For starters, Google the search phrase "Facebook and child abuse". Like AOL in its heyday, Facebook is the primary communications tool for the scum of the earth. U.S. law enforcement allowed AOL to get away with it and they're allowing Facebook to get away with it too. Why? Your guess is as good as mine.

As if that weren't bad enough, Facebook, using carefully calibrated algorithms, targets psychologically vulnerable members of society, especially children, with a stream of negative content knowing that it causes "greater engagement" and thus more profits for the company. This analysis comes straight from Sean Parker the founding president of the company. Facebook/Instagram is an ideologically-based censorship and propaganda platform, at least as bad as Google and in many ways worse (which is an accomplishment.) In addition to all this, the platform does a first-class job of un-training people how to read, while diminishing their attention span and their ability to interact with the live human beings around them.[44]

What, if anything, is to be done about these companies?

All three of them are monopolies that violate U.S. law. If the Justice Department and FTC were anything other than political organizations, they'd break them up. If the companies in question were to suffer in the process, it would be no great loss.

---

44   "Non-informing, knowledge-eroding medium." I don't know who coined this phrase about Facebook, but it's very accurate.

The model of getting as big as you can as fast as you can, flim-flaming and outrunning regulators, and developing a massive lobbying (i.e. bribery) war chest to defend the resulting abusive monopoly you created is much closer to fascism than it is to free enterprise. Benito Mussolini, who coined the term fascism, believed in the state as the ultimate end and the fusion of select corporations with the state. Enabling and, in some documented cases, collaborating with the aforementioned corporations in their abuse of the public is part of a long list of ways our governments have let us down.

To paraphrase George Carlin, the people at the top of today's Internet food chain don't care about you. Their sole focus is how to turn the network into something that pays them or gratifies other desires like having political power. If that comes at society's expense, so be it.

The good news, and there is good news, is that despite their best efforts and their limited success in turning large swaths of the Internet into the equivalent of a toxic waste dump for their own profit, there is, for now at least, still plenty of room for the good people of the world to meet each other, learn new things, collaborate, and add to the world's storehouse of knowledge. Just don't be entirely dependent on Big Tech to do so.

Like Big Pharma, defense contractors, big banks, and other corporate parasites, Big Tech is not your friend. Not so coincidentally, big Internet platforms derive a significant portion of their income from defense contracting and are simultaneously attempting to fuse with the pharmaceutical industry. See Google's massive pharma investments and accompanying information-suppression operations and Amazon's attempt to turn itself into America's corner drugstore.

Therefore, the ongoing imperative is to use Big Tech, and not be used by it.

Related to this, never wait for the government, which now includes academia and the news media, to do the right thing for you. They're incapable of it. The sooner you realize this, the less time you'll waste and the less frustration you'll experience. You are the person most likely to improve your own life and your community and by extension your country and world. There are good sources of information on the Internet about every subject you'll ever want to know about and every problem you'll ever face.

There are kind, capable people in this world and if you look for them, you will find them. Remember to appreciate them and be one of them. Don't let anyone label you as "less than" because of your current economic or career status and don't label yourself that way either. The secrets of life are learning how to work, how not to give up, how to learn, and how to get along with people. These are hard lessons for everyone, but we're all well-built to figure them out, and it's a process that never ends.

Take time every day to witness and get to know the Creation around you, which despite all efforts to despoil it, is still beautiful, rich beyond words, and always rewards respectful study. Nature is the baseline reality and you are one of its treasures. The more you learn, the more you'll realize that this is true. Finally, as a good rule of thumb, the less time you spend in front of screens, the happier and more fulfilled your life will be. The Internet was created to enhance your life, not to be a substitute for living.

- Part Five -

Collected Writings and Talks

Appendix I:

Why We Will Never Have Interactive Television — and Why You Should Stop @#%*-ing Around and Get on the Internet NOW!

DM News
August, 1994

When I look at the plans big corporations have for Interactive TV (over 30 prototypes have failed in tests so far), I'm reminded of the opening of Claude Hopkins' "Things Too Costly" chapter in the 1923 book "Scientific Advertising."

Hopkins wrote: "Many things are possible in advertising which are too costly to attempt . . . changing people's habits is very expensive. To sell shaving soap to the peasants of Russia one would first need to change their beard-wearing habits. The cost would be excessive. Yet countless advertisers try to do things almost as impossible. Just because questions are not ably considered and results are untraced and unknown."

Television watching is not an interactive activity. Yes, there will always be a small handful of enthusiasts buying a new gizmo and playing with it for a little while, but in all the years that interactive TV tests have been run, none of these systems has inspired passion or, more importantly for our purposes as direct marketers, habitual use.

Like the peasants of old Russia, TV watchers are set in their ways. Yes, they'll get up from the sofa to order a pizza or a magazine subscription. And companies like QVC have discovered a unique subspecies: consumers who watch TV with their telephones and credit cards on their laps, the better to order jewelry, clothing, and assorted tchotchkes. While it is possible to not only make money but also become rich selling to this group, QVC's audience is not the future of electronic marketing.

We will have an information superhighway, but it will be based not on the well-named "idiot box," but on networked personal computers.

Some evidence: Bill Gates, the world's most important software man, and Andy Grove, the world's most important chip man, say so. Software sales already dwarf the output of film, TV, and recording studios. Dollars spent on personal computers exceed dollars spent on TVs and the gap is widening exponentially. And what's most important is that these two giants, and the millions of industry members they represent know interactivity like Claude Hopkins knew salesmanship in print. They're not speculating on something that might someday be. They're observing a phenomenon in progress.

Not convinced? How about this bizarre but unmistakable sign that real people are beginning to embrace the idea of PCs as a home appliance: Howard Stern recently took time from his usual on-air banter to explain how he's set up his laptop computer to operate his compact disc player. Then he went on to extol IBM's OS/2 operating system, adding that Microsoft Windows "sucks" by comparison. Not a commercial. Just the most influential radio personality having a chat with an audience of millions he knows like the back of his hand.

Still not convinced? How about a little history. Let's look at the telephone, the wildly successful interactive communications medium that people use several billion times a day. Did AT&T decree that every home would have a telephone and that they would be used to order products and information? No, far from it.

Although patented in 1876, it wasn't until the early 1920s that the telephone reached the same level of in-home penetration the personal computer enjoys today. At first, telephones were expensive, unreliable, hard-to-manage devices. They were used by professionals like stock traders, journalists, and law enforcement officers who had an obvious and immediate use for them. Others, authorities like the chief engineer of the British Postal and Telegraph System (then the biggest telecommunications infrastructure in the world), scorned them as an impractical, unnecessary luxury. He said of his telephone: "I have one in my office, but mostly for show. If I want to send a message, I employ a boy to take it."

In a very real sense, telephone users invented the telephone as we know it today. If you think about it, how could it have happened any other way?

The inventors of new technologies are notoriously shortsighted when it comes to envisioning how their creations will be used. Alexander Graham Bell thought the telephone would best be used to pipe music into people's homes. No one anticipated the telephone becoming an instrument for personal conversation or imagined the thousands of unique businesses that would spring up around them. The telephone may have been born in a laboratory and financed on Wall Street, but it was regular people in the real, everyday world who invented practical interactivity.

Given this history of the telephone, does it really seem likely that a hodgepodge of publishers, broadcasters, cable operators, and telephone service providers are going to be able to create, out of whole cloth, an interactive communications system that real people are going to use and pay for? It's an attractive fantasy, but that's all it is. The hard-for-some-to-swallow fact is the "golden age" of three TV networks and one telephone company ruling the world and making all the money is over for good. No amount of wishful thinking, research dollars, or hype is going to bring it back in a new incarnation. Interactive TV, like the emperor from the fairy tale, is buck naked.

What is happening is that while millions of corporate dollars are being spent trying to cook up a workable interactive TV scheme, millions of people are beginning to use their once "expensive, unreliable, hard-to-manage" personal computers for personal communications. And a small but growing percentage are starting to fool around with sending and receiving video. The importance of these twin phenomena cannot be overestimated.

Just as the British Empire's leading authority on telecommunications dismissed the telephone, you may choose to dismiss these new developments as an extravagant fad. You have every right and you'll be in good company. Less than 20 years ago, the CEO of Digital Equipment, then the world's second-largest computer manufacturer, ridiculed the very notion of personal computing. Today, the company is gushing barrels of red ink with no end in sight.

The fact is, after an appalling start (few industries have gotten away with such a shameless run of empty hype and broken promises), personal computers are getting cheaper and faster. The ease with which people can communicate via their computers is increasing rapidly. And business people are discovering the wonders of E-mail. If you're already an E-mail user you don't need to be sold, but if you're not you're missing out on something significant.

I first got hip to E-mail when I was trying to track down an acquaintance from college I hadn't spoken with in 10 years. No one I knew had a current address or telephone number for him, but someone did have an Internet address. I called up another friend who had an Internet account and asked him to send David a message to call me. A day later, the phone rang and we were connected. In the course of our conversation, David said something I found startling:

"It's a good thing you sent me an E-mail. If you'd written or phoned me there's no telling when I'd have gotten back to you. I barely look at my mail and I'm terrible about returning phone calls."

Here was something I had never encountered before in my life and didn't even suspect existed; a well-paid person with ample disposable income who prefers E-mail to all other forms of mediated communication. Not much later, I read that Bill Gates, America's richest man, routinely receives and responds to E-mail messages from total strangers. Intriguing to say the least. What kind of phenomenal direct mail or telemarketing strategy would you have to employ to accomplish this same feat?

Now that I'm an Internet user, I'm as fond of it as my friend David is. I can answer seven or more E-mail messages in the time it takes to answer one telephone message.

Contrast the following all-too-common telephone scenario with E-mail communication. Let's imagine you want to alert a colleague to an article in this week's DM News.

(Ring, Ring, Ring)

Receptionist: *Hello, Acme, Zenith, and Timeblaster. Please hold.*

(Hold for 10 seconds.)

Receptionist: *Hello, How may I help you?*

You: Yes. *Is Bob Smith in please?*

Receptionist: *No, he's out of the office right now. His calls are being switched to me. Can I take a message? Oh, could you hold please?*

(Hold for 10 seconds.)

Receptionist: *Thanks for holding. Can I take a message for Mr. Smith?*

You: *Yes. This is Ken McCarthy and I'd like to tell him...*

Receptionist: *Is that "M-C" or "M-A-C"?*

And so on, round and round the mulberry bush, just to transmit the simplest of messages.

Now look at how Internet E-mail handles this:

1. Turn on your mail program.

2. Address E-mail to Bob Smith by typing "bob.smith@acme.com" or something similar and hit return (if his address is already on your computer, this can be done with just one keystroke).

3. Write in the topic: "DM News article" and hit return.

4. Say your piece: "Read a great article on E-mail in this week's DM News. Check it out. It's on page 40. Ken." Hit return.

The message is automatically sent and is in his mailbox, just the way you sent it, in a matter of seconds. If he's on the other side of the world, it might take a bit longer depending on traffic, but there's no charge for the distance sent. You pay for your local Internet connection only.

When Bob receives your thoughtful message, he can respond in a few seconds just by hitting the reply key and typing "Thanks." The E-mail is pre-addressed to you, the sender, for easy response. Bob can also quote your message for reference with a keystroke to refresh your memory about what he's responding to.

For quick, informal messages, the kind that everyday business depends on, which system would you rather use? And I've described only one of many ways to use E-mail. E-mail users are inventing new ones all the time.

You have a mailing address, a telephone number, and a fax number, don't you?

Now you need one more line on your business card; your Internet E-mail address. This simple step will open a new channel of communication between you, your company, and an audience of what some estimate to be as many as 50 million well-educated, affluent people worldwide. Your prospects and customers can use E-mail to make inquiries, order products, and offer suggestions. And once you learn a bit about the unique social niceties that have grown up around E-mail, you can use it to sell.

Appendix II:
Which Way the Web?
Pac Bell Center, San Francisco, CA.
Talk - November 5, 1994

"We're here today because this year the Internet has really changed. It's gone from being a "network of networks," whatever that means, to being a medium. A true medium.

How did that happen? Well, Mosaic is the reason it happened. And the Internet now takes its place with television and radio and publishing and CD-ROMs as a medium, not just a place for computer people to hang out. The numbers of people on the Internet are very small relative to those other media. But that's not the point. That's not what defines a medium. It's not just numbers.

How did this particular meeting come to be? I was amazed to discover when I did some inquiries last spring, that very few multimedia producers are even on the Internet, let alone Internet savvy. It's a small percentage. It's a growing number. But it's less than 20%. And I thought that's very odd. I mean, here we are in San Francisco. It's the world's center for multimedia development. The Internet, of course, by definition is distributed all over the world. But a lot of the great Internet talent is right here in the Bay Area. These people are literally neighbors, and they hardly talk to each other. Kind of strange. It's particularly strange because the Internet really needs multimedia developers.

What's going to determine whether the Internet succeeds or not is not technical issues, it's going to be content issues. Is the programming that's going to be on the Internet interesting enough, motivating enough, and enlightening enough that people are going to want to tune in and use it? That's going to be purely a content issue.

Nobody goes to the movies to watch the technology of the movies, they go to the movies for the story. When we think about movies, we think about the Academy Awards. 100 million-plus people watch the Academy Awards every year. How many people know about or think about the annual SimT convention? A very, very small number. And that's what's going to happen on the Internet too.

So we need content. And who better to produce digital, interactive content than multimedia title producers? They're the only people on the planet that know how to do it, that even think about it.

But there's also a reason why multimedia people should get hip to the Internet very fast. And that is because the CD-ROM business, in my opinion, as a person from the publishing business, is a lousy business to be in. It's terrible. Why? Well, let's say you spend $100,000 to produce your title. Okay, so you've got your work. Now you've got to press it. You've got to package it, and the packaging often costs a lot more than the pressing. You've got to inventory that stuff. So, it's like taking a big pile of money and putting it in a closet for a while. Not much fun.

You've got to find a distributor. You've got to beg a distributor to take your material. Then you've got to give them a big piece of the sales price. Then your distributor has to persuade a retail store to take the disc, to take your title. Then you've got to persuade a $ 6-an-hour clerk to take those titles out of the back of the room and put them on the shelf, and that can be sometimes the hardest job of all. And it's one that by definition you just can't do because there are so many stores and so many people to deal with.

But as bad as that is, there's an even more important reason why CD-ROMs are not a great deal for the producer. You have no contact with your customer; you have no relationship, okay? Their relationship is with the store or the catalog that they're buying it from. So, you've gone through all this effort to produce this title, to excite somebody about your work, and you're not there to actually be with them and sell them - and the most important thing is to sell them additional products. You have to go right through the old channel of distributors and stores all over again.

Now, the Internet, the thing that excites me about it, is it allows you direct contact with your customers. No middlemen. You produce it, you distribute it. And you can build up a following and profit from that following.

One of the sorts of tragedies of the way our media system is set up so far, is we all have to go through movie companies, film studios, recording companies, or publishers to get our work done. And they don't make their decisions based on quality. They just don't. Their decisions are made on kind of a lifeboat basis. "We have so much money to develop this title, whether it's a book or a movie or a record, it's going to cost this amount of production costs. The distribution costs are phenomenal. So we've got to budget those in advance, and the marketing costs are going to be phenomenal. And we have limited resources."

So, what's getting produced these days is not necessarily the best, or what's best for society, or even when people are going to benefit from or be interested in. It's what fits a certain parameter; what fits in the lifeboat.

The Internet can do a lot to change that. And we've seen some success stories already. So, I went to the multimedia people, and I said, "Would you come to a meeting about the Internet?" And they said, "Yeah, we'd love to come." So, we have a lot of multimedia title developers in the audience. Then I went to all my filmmaking friends and said, "You know, you should get in on this too, because one of your problems is distribution. And while we can't distribute your independent films on the Internet, today, what we can do is we can distribute your trailers." And as anybody knows, that's the ad for the movie, that little 30-second, 60-second clip. That's enough to sell a movie. That's enough to raise a million dollars, or $5 million, or $10 million to produce a movie.

When I worked in the movie industry in New York, the procedure for doing an independent film is you make your trailer first. And you take the trailer around to all the investors. They get all excited about it, they give you the money, and then you make the movie. So right now, any filmmaker with the will, who can find a good Internet person to work with, could put their trailer on the Internet, make it available globally, and begin to attract an audience for his movie or her movie before it's even made. That's very exciting. And that's why NAMAC, the National Association of Media Arts and Culture, is a sponsor of this seminar; they represent hundreds, probably 1000s of independent filmmakers. And this is their chance to get those filmmakers out into the world faster.

Then I thought about record companies. And of course, they have the same problems as book publishers. And then I thought about authors and publishers of print, and invited them too. And I also thought about the advertising business. The advertising industry makes more movies than Hollywood. Commercials shoot more film than Hollywood and are responsible for more printing than the magazine and book publishing industry. Probably does more recording than the record industry.

So eventually, when the Internet really matures, the people that are going to be on it the most, for better or worse, are people from Madison Avenue. And they're going to be the ones with the money and the resources to do all sorts of amazing projects. I invited some of them to come as well today. So that's who is here and why and what we're about.

What I'd like to talk about in my remarks are the stories that have been missed this year by the media. The media has done a great job of hyping the Internet and getting a lot of people interested in it and a lot of people excited about it. But they've told a few stories slightly incorrectly, or in a confusing manner, or have left certain things out. Since I've got a podium for a half-hour, I'm going to do what I've always wanted to do: correct the newspaper.

First misconception: A lot of people are talking about cyberspace, and the information superhighway, and this idea that we're going to create this alternate environment that we're all going to live on, and everything's going to be done there, and that the measure of our success will be that everything is done there, and we should be gearing all our attention to create this place that's completely independent of the rest of the world.

That's crazy. That's just absolutely insane. Would you start a business that only did business on the telephone? In other words, you wouldn't have a store, you wouldn't talk to anybody in person, you wouldn't send mail or receive mail, and you would only be on the telephone? Of course not. Would you have a business that only had a store, but didn't use a telephone and didn't use the mail system? Of course not.

The picture of every mature business is that they use every conceivable channel available to them - they use the mail intelligently, they use the phone system intelligently, they use video intelligently, and they use the Internet and cyberspace intelligently. So let's get rid of this idea that we're trying to create some alternate world that's going to be completely independent of all the other media that exist.

What we're really going to try to do is find a place for the Internet amongst all these other existing medias - to integrate, and let the different media support and coordinate with each other.

The second crazy thing that I hear going around a lot is "How are people going to find out about what's on the Internet?" You know, how many places can we post on the Internet to tell people about what we're doing?

Well, how do people find out about your telephone number? And how do people find out the location of your store? You advertise. And you use every available means - you use television commercials, or radio commercials, or direct mail campaigns, or space ads in magazines, or you put on conferences or T-shirts, or all the other things that we do to get people to dial our telephone.

These are the things that we'll do to get people to dial-up our website. Right? So we don't have to worry that there are not enough advertising opportunities on the Internet itself to get people to come to our site. We just have to look out at all the other medias that are available and use those to drive people to our site. Make sense?

The other story that the media I think has gotten slightly wrong, is more of a missed story. And this has to do with the computer bulletin board phenomenon. There are currently, according to Jack Rickard the publisher of Boardwatch Magazine, somewhere between 50,000 and 70,000 computer bulletin boards in operation in the United States today. And those are public boards that regular people are setting up that don't count corporate boards, government boards, or any of those things.

What's the bulletin board? Well, it's a mini America Online or mini CompuServe. A guy and maybe some of his friends set up a PC, they buy some BBS software, they buy some phone lines, and some modems, and they've got an online service. And there are 50,000 people out there doing this, 50,000 plus. That's the story and it's not really been covered.

But the big story that the Internet press missed this year is that this summer the bulletin board software makers have announced products that will allow these little bulletin boards to become Internet service providers. If you've got 50,000 to 70,000 entities, which probably represent maybe 150,000 or 200,000 technically knowledgeable people - some of them do have business savvy and have been able to build good businesses with limited resources of computer bulletin boards. Now, suddenly, overnight, they're going to be able to provide Internet accounts in all kinds of places that, so far, we haven't been able to get Internet access to for regular people.

So there's going to be a flood of Internet services in the next year or two. And Jack, who's a pretty sober-minded individual, thinks that there are going to be an additional 10 million slip accounts opened up on the Internet in the next year. Why? Because we're going to be able to do it, and we weren't able to do it before. We're going to be able to do it at a cost that's reasonable.

Unlike some of the larger services where you call and you get a busy signal or a recorded message and you can never actually talk to somebody to help you, the nice thing about a small computer bulletin board - these businesses are scaled to serve 500 customers or 1,000 customers. So there are all sorts of possibilities of customer training and customer service that companies that I won't mention, aren't able to provide or have ceased to provide.

And it's a neat little business. Imagine if you had 1000 subscribers, which you could build up over a year or two which wouldn't be hard. And you're charging $20 a month. That's $20,000 a month for a business that maybe needs two or three or four people maximum to operate. That's not a bad little mom-and-pop type business. And you might be asking, "Well, you know, is it really realistic that all these disorganized, scattered, tinkerer computer-type people are going to be able to create a large media?" And the answer is they've done it once they can do it again.

Here is a magazine that was published in 1925, called Radio Broadcast (see Figure 1). It's an interesting thing to look at. First are the titles of what's being discussed. Number one, a good four-tube receiver, you know. I mean, how many people listening to the radio today have any knowledge or interest at all in what's going on inside their radio? They don't care. That's how the Internet should be and will be.

Next topic, choosing a B battery eliminator. And this was important for people who were listening to the radio 70 years ago, or so. And then finally, the million-dollar question - this has never been a problem before - who is to pay for broadcasting, and how? Right? Sound familiar? Well, we worked it out somehow. And we're certainly going to work it out on the Internet.

Just some funny things about this picture.

I mean, this could be a PC guy, right? No problem. He's got his manual open on the floor, actually a pile of them, desperately trying to figure out what's going on. Batteries sitting behind his chair, and then a whole tangle of wires and headphones that no longer work, but might work again someday and could be useful for something. He's smoking a pipe, and I'll leave that up to your imagination. And he's extremely excited - can you see the expression on his face? I mean, you know, he's wired. 70 years ago. And that turned into broadcasting, a multi-multi-billion dollar business.

The neat thing about the Internet is that broadcasting was co-opted by large companies because of the nature of the technology. You needed millions of dollars to set up transmitters and so on. The Internet is not that kind of

Figure 1
Radio Broadcast Magazine cover, March 1925
"Who Is To Pay for Broadcasting and How?"

an animal. I think it can stay distributed and can stay an opportunity for small businesses and individual artists.

Okay, what else have we missed in terms of stories? Oh, good - demographics. You know, who's on the Internet and who's not on the Internet? There was a story, I think it was in the *Times* or the *Wall Street Journal* recently that said, "Well, the people on the Internet have more time than money." Well, hey, surprise, that defines 95% of the world.

I know big companies and the Wall Street Journal and publications like that, like demographic studies, but they make no sense in an explosion of a new medium. What sense would it have made to do a demographic study of television owners in 1949? There are 8000 of them. Let's say you've done a brilliant study, and you've just wrung every answer date out of that - what would it have proven? Nothing.

What if you studied PC owners in 1978? That would have told you a lot. Yes, when an industry is mature, if you're trying to figure out whether you should buy an ad in *Steelmaking Today*, then demographics are really important. But with a medium that's growing, 20% a month or more, that's not the point.

The question is: Is this a real medium? Is this something that's really going to last? Is it really going to grow? And the answer to that is: does it fill a need? And the answer to that question, I think, is: yes. And that's why it is growing so fast. So, disregard all demographic studies regarding the Internet. I don't see what the point is in them, at this point. Okay.

There was another article in another major publication saying that "Well, people are setting up Internet catalogs, but nobody's buying anything." Did anyone see that story? Well, I've had a bit of dealing in the direct marketing industry, which includes direct mail, and catalogs, and producing infomercials, and direct response television commercials. And I'll tell you right now, 19 out of 20 direct marketing ventures in the old-fashioned media of television and print don't work either. It's quite hard to create a direct marketing business that works. So it shouldn't be a surprise that some of the early pioneers of Internet cataloging might not be getting the sales that they had initially hoped.

Number one, the market is a little thin. While there are millions and millions of people with some kind of Internet access, not all of them know how to find catalogs. And number two, a lot of the people that are running online catalogs are primarily engineers, they're not marketers.

And take this on faith: one of the hardest businesses to run, from a direct marketing point of view, is a catalog business. It's a very brutal business. The margins are razor-thin. It's tough to sell things at a distance. So, it shouldn't be a surprise that the initial attempts at selling the Internet are running into certain difficulties.

Bandwidth. One limitation I always hear about is bandwidth limitations. And I'd like to give you an analogy. Let's say it's 1880. And we're down at the telegraph station, and the train is pulling in, and you point to that train, you tell someone, "Someday we're going to take that train, and we're going to shrink it down, we're going to put rubber wheels on it, and we're going to create a road system so that you can take that train anywhere you want to go, and it's going to be so cheap that anyone that really wants to have one can have their own train, they can go drive it anywhere they want." What do you think the reaction would have been? "You're nuts. You've been taking too many of those, you know, opium-laced patent medicines that they're advertising in the back of the magazines."

Let's say you walked into the telegraph office - and what was the telegraph office? It was a line of people waiting patiently to hand their message to a technologist, who would, using some binary code, send a message to another technologist who would translate it, and then pass it off to somebody. Let's say you went up to that person and said, "Someday you're going to have your own telegraph office, it's going to be in your house, and you won't need anybody to operate it for you because you're going to be able to talk to anyone you want to talk to who's got a line." Again, there's going to be the same reaction.

We're capable as a race, as a human race. And I think Americans, in particular, if I can be a little prejudiced, are great at creating all sorts of amazing leaps of technology. We're really just talking about adding a little bit more bandwidth. We're not talking about inventing something new or laying the first transatlantic cable - which was quite a difficult physical feat.

We're talking about taking technology that we already have, figuring out how to pay for it, and just installing it. So the bandwidth problems...When they'll be solved, I don't know, but they're coming. And about bandwidth. Multimedia developers may say, "Well, I can't distribute my CD-ROM on the Internet. Why even bother?" Well, create a product that works on the existing network as it is. The people that produce Doom seem to be doing all right with that strategy. 500,000 orders are already sitting on their desks back wherever they are in Texas, for the first store distribution.

There's a principle in direct marketing that if you can't get somebody to distribute your product in a store, you run your own ads, you run your own mailer, you run your own infomercial. So, you run your own direct response and TV commercials, and that forces the larger distribution networks to take you seriously and adopt your product. I think the Internet is a way to force distribution. I would look at it that way if I were you.

Okay. Last story. This is a really current one.

Steve Case (CEO of America Online) recently got in trouble for doing what? Offering to rent his mailing list. Why would he do such a thing? Why would he violate the privacy of his customers?

Well, first of all, everybody you do business with keeps your name. And everybody that's smart, tries to leverage that name in some way, either by renting the name to another business, co-venturing with another business, or taking a new product line on and attempting to sell you that product. It's the oldest thing in the world that's not computer-based. It goes back to the beginning of the century with the distribution, trading, and cultivation of mailing lists.

But why would Steve Case want to do this? Well, here's some math. You have 1000 names, the rental price for 1000 names is approximately $100. Okay, that's about what I imagined America Online could get per 1000 names rented one time. Conceivably, he could get more. Okay, so he has a million members, right? So that's 1000, 1000s. So, one rental of his list is 1000 times $100, which is $100,000, which is pretty much straight to his bottom line.

What's involved? You take the tape, you put it in the mail, and it goes to the mail house. And that's it. So 20 bucks, 50 bucks. Now there are brokers involved. And the brokers who arranged these deals take 20%. But still, $100,000 per turn is not bad. How many times can Steve run his list in a year? Well, there are aggressive companies that rent their lists 30 to 40 times a year. So you have to take $100,000 and multiply it let's say by 30. That's $3 million. Not bad, okay, but it gets even better.

Steve doesn't have a million names. He's got 5 million names, or close to that. Because there are people that have signed up, given their information, and then not continued with the service. It's not the number I just said which was 3 million. It's 5 times 3 million. Sitting in the back offices of America Online is an asset worth pretty much net $15 million a year.

The question is not why did Steve Case try to do this – it's what took him so long? And this is where I'm going to end. What took him so long is that he was coming from a technology – he is a marketer, originally – he's coming from a technology background. And he just simply was not aware of the income possible from list rental. Why? Because this whole industry of providing online services is so new, a lot of the old tricks that the print publishers have learned over the last 100 years, of how to squeeze every possible dollar out of the situation, they just haven't learned.

You could make money in the online services business up until this time right now simply by putting up a service because you would be the only one. A lot of people say, "This is a lot like Gutenberg's Bible, it's like the first printing press." I don't think that's an accurate analogy. I think Gutenberg's Bible was more like the Harvard Mark One, I forget the name of it, but that original computer they made. That's what Gutenberg Printing Press was like, In reality. There weren't that many of them. The vast majority of people could not read. Books were still the property of popes and cardinals and kings and princes. The more accurate analogy is the late 19th century.

What happened in the late 19th century? Because of a coming together of a lot of different things, all of a sudden print became very cheap. We see print everywhere, and we assume that it's always been ubiquitous. And we assume that newspapers and magazines have always been around. The fact of the matter is, that's not true. We did not really have an explosion of print and print as a mass medium until post-Civil War.

For example, in 1850, there were 254 newspapers in the United States in total. 50 years later, there were 2,600 daily newspapers, 520 Sunday newspapers, and 15,500 weekly newspapers. Before the Civil War, most families were lucky to have a single Bible, and it was a family Bible. And it was passed down from generation to generation. After the Civil War, you had a thing called the Sears catalog in which a company very intelligently realized, "We've got this printing deal that's really cheap, we've got this postal system that's incredible. We've got this railroad system. Let's give out some amazing free thing - The Sears catalog - put it in everyone's home and create a vast selling tool with that." But that couldn't happen till after the Civil War.

And the last thing I want to say, look at the old-timers, look at the old media. For instance, we think America Online is such a hot thing because you can get electronic mail, you can play games, and you can chat with people. And you can read the news, right? All these amazing things. 100 years ago, in the country store, you got your mail, you read the news, you chatted about things, you played checkers on the cheese barrel, whatever. We're basically just recreating something that we already had, in electronic form.

The other thing that I would recommend is that you think about the Sears catalog, think about the idea in 1885 of giving away this marvelous print thing that was so intrinsically interesting that people loved just to page through it and happened to buy millions and millions of goods as a result of paging through this product. The Internet makes the distribution of digital goodies extremely easy.

The last thing that I want to leave you with was the big story that everyone missed this year. There was an invention 150 years ago that made railroad travel possible. It made mass production possible. It forced the creation of this thing called standardized time and wristwatches and pocket watches. It was the precursor of the telephone. It created recorded music. It created broadcasting. One invention - all those things flowed directly out of this one invention. Does anyone know what it is? 150th anniversary this year? What's that?

Telegraph. So many amazing things came out of this. A lot of people don't know that recorded music came from the telegraph. Edison was not trying to record music. He was trying to make a way to automatically relay telegraph messages. That little wax cylinder was meant to record dots and dashes. You're going to pull that cylinder out, stick it into a machine, and it would replay the dots and dashes so that the message didn't have to be hand relayed. The telephone was originally an attempt to have many, many messages, telegraph messages on a single line, and they discovered "Wow, we can actually put a voice through this as well." And on and on. You couldn't have high-speed train travel until you had a telegraph, because you couldn't send a train barreling down a track at 60 miles an hour unless you knew with some degree of certainty what was going on about a half-hour away or an hour away.

And mass production came directly from Chicago. The meat industry realized, "Hey, we could take all these cows, put them on a train, feed them all in one place, dress them, and send them out all at once. Henry Ford didn't invent mass production. Henry Ford studied with Richard Sears, he went to his catalog company and saw how Richard Sears handled tens of thousands of orders a day. And Richard Sears went to the stockyards in the slaughterhouses of Chicago, and saw how mass production worked.

So the telegraph really produced just about everything that we know today. And this year is its 150th anniversary. I think it's fitting that this is the first year that the Internet can really be looked at as not just a collection of networks but actually a medium unto itself, that could probably produce as many or maybe more changes in our life."

The video recording of this talk can be seen at: www.HowtheWebWon.com

Appendix III:

The Power of the Electronic Press Unleashed — An Interview with John McLaughlin

Net Ventures Vol.1 No.1
May-June 1995

"Right now the Internet is like a printing press with paper and ink next to it. Everybody's standing around saying what a great invention it is. OK, fair enough, but it's time we actually started printing."

— John McLaughlin: Publisher of SunFlash

Welcome to the first issue of Net Ventures, the journal about how people use the Internet to solve problems and create opportunities for themselves, their customers, and the organizations they work for.

Trying to learn about the Internet from reading books and attending conferences alone is bound to be a frustrating task. There are certain things you simply can't learn from books. Ask 100 Internet pros how they got their knowledge of the Net and invariably you'll get the same answer: someone personally showed them the ropes and they took it from there.

NET VENTURES

Vol. 1 No. 1   In Depth Profiles of Internet-based Business Ventures   May-June 1995

John McLaughlin: The Power of the Electronic Press - Unleashed

> "Right now the Internet is like a printing press with paper and ink next to it. Everybody's standing around saying what a great invention it is. OK, fair enough, but it's time we actually started printing."
>
> **John McLaughlin: Publisher of SunFlash**

Welcome to the first issue of Net Ventures, the journal about how people use the Internet to solve problems and create opportunities for themselves, their customers, and the organizations they work for.

Trying to learn about the Internet from reading books and attending conferences alone is bound to be a frustrating task. There are certain things you simply can't learn from books. Ask 100 Internet pros how they got their knowledge of the 'Net and invariably you'll get the same answer: someone personally showed them the ropes and they took it from there.

Without stepping on any regional toes, I think it's fair to say the biggest proportion of the world's Internet experts live right here in Northern California. And, if you wait long enough, you can count on the rest of them passing through here at one point or another.

It occurred to me one evening, while having dinner with an Internet friend, that, as an Internet-interested San Franciscan, I regularly enjoy what people in other parts of the world can only dream about: easy access to an incredible superabundance of Internet talent.

Then I played "What if." What if instead of limiting my conversations to me and whoever else was in earshot, I taped them? And what if I found somebody to edit them so they made sense? Internet people are amongst the most free-handed people I know with their time and knowledge. Why not multiply the benefits of their generosity so that thousands of people can make progress together?

Let me reveal my prejudices up front. While I am excited about what the Internet can do, I am much more interested in what people do with it. I was
(Continued on back page)

INTERVIEW

J: John McLaughlin   K: Ken McCarthy

K: We're here with John McLaughlin of Sun Microsystems. You work in the Florida office?

J: Right. I'm a Systems Engineer, and I also publish electronic newsletters.

K: And the name of the newsletter you publish is...?

J: ... it's called SunFlash, and I've been producing it for seventy-one months. See, it's seventy one months because they're numbered by volume. So, I'm just coming up on our sixth year.

K: What's your subscription base? How many people read your publication?

J: It's about 140 or 150,000. It's really broad.

K: How do you manage a subscription base that is so large?

J: The top level that I post has about 5,000 entries, of which at least 1,000 are aliases to other locations, either other Sun offices or our larger customer sites which have their own redistribution offices.

K: In other words, you send the 5,000 and at least 1,000 of those people, in turn, send it to ten, 100 or 1,000?

J: Exactly. I know of two or three locations that I've talked to over the years that have in the order of 1,000 individual subscribers.

K: Wow.

J: The reach is actually a little bit greater than that. I know of a number of sites, including some of my own local customers, that have about 1,000 workstations with only about 20 people subscribing and those 20 people are kind of the information technology gatekeepers. They forward the proper messages to the proper people. If it's related to networking they forward it to the networking group. That way the networking group doesn't have to monitor the 100

Front page of Net Ventures Vol.1 No.1
May-June 1995

Without stepping on any regional toes, I think it's fair to say the biggest proportion of the world's Internet experts live right here in Northern California. And, if you wait long enough, you can count on the rest of them passing through here at one point or another.

It occurred to me one evening, while having dinner with an Internet friend, that, as an Internet-interested San Franciscan, I regularly enjoy what people in other parts of the world can only dream about. easy access to an incredible superabundance of Internet talent.

Then I played "What if?" What if instead of limiting my conversations to me and whoever else was in earshot, I taped them? And what if I found somebody to edit them so they made sense? Internet people are amongst the most free-handed people I know with their time and knowledge. Why not multiply the benefits of their generosity so that thousands of people can make progress together?

Let me reveal my prejudices up front. While I am excited about what the Internet can do, I am much more interested in what people do with it. I was struck by an interview in "Morph's Outpost." Marc Andreessen said something like "The part of the Internet I use the most is e-mail. Otherwise, I haven't found that many things on it that are useful."

I couldn't agree more. The tools are impressive and pregnant with potential, but we have barely scratched the surface of discovering how to use them and how to make them pay.

John McLaughlin's work with electronic publishing is the perfect subject to launch this publication. He has accomplished things that most publishing professionals would tell you are impossible.

Working alone in his spare time (he works as a field engineer for Sun Microsystems), John publishes a periodical that is read by at least 10% of the members of a $25 billion industry. And he does it without ever seeing a printing bill or visiting the post office!

Has John become rich from his efforts? Not yet, but he didn't start out to become rich. He was merely trying to create a service that would help his customers. Six years later, he's developed a simple, yet incredibly powerful publishing machine that is worth its weight in gold. After reading this interview, you'll see for yourself why the Internet has such an enormous potential for bringing companies in closer contact with their customers.

Next issue: Software.net

***

INTERVIEW

(**J**: John McLaughlin; **K**: Ken McCarthy)

K: We're here with John McLaughlin of Sun Microsystems. You work in the Florida office?

J: Right. I'm a Systems Engineer, and I also publish electronic newsletters.

K: And the name of the newsletter you publish is...?

J: …. it's called SunFlash, and I've been producing it for seventy-one months. See, it's seventy-one months because they're numbered by volume. So, I'm just coming up on our sixth year.

K: What's your subscription base? How many people read your publication?

J: It's about 140 or 150,000. It's really broad.

K: How do you manage a subscription base that is so large?

J: The top level that I post has about 5,000 entries, of which at least 1,000 are aliases to other locations, either other Sun offices or our larger customer sites which have their own redistribution offices.

K: In other words, you send the 5,000 and at least 1,000 of those people, in turn, send it to ten, 100, or 1,000?

J: Exactly. I know of two or three locations that I've talked to over the years that have in the order of 1,000 individual subscribers.

K: Wow.

J: The reach is actually a little bit greater than that. I know of a number of sites, including some of my own local customers, that have about 1,000 workstations with only about 20 people subscribing and those 20 people are kind of the information technology gatekeepers. They forward the proper messages to the proper people. If it's related to networking they forward it to the networking group. That way the networking group doesn't have to monitor the 100 or so articles that I produce each month.

K: Is SunFlash published monthly?

J: No, I produce continuously. Sun, for example, publishes press releases which I try to turn around immediately. I try to stabilize 20 or so articles each week. What I discovered was that if I give myself one deadline each month, it would be so much work I just couldn't do it. So, by doing it as the data became available it's manageable. At my busiest, I wait until the weekend to catch up.

K: With that kind of flexibility, do some people like to subscribe on a daily basis and others weekly?

J: Some people do, so I created a weekly and a monthly version of article summaries. If they wish to get the full article, they can send the article number to an automatic response program which returns the full text of that article. That way they can scan 25 articles and send for the four or five they're interested in and they will arrive ten or twenty minutes later.

K: What kind of software allows you to automate this process so efficiently?

J: Many in my audience are Sun users and use some Sun-specific tools that are shipped with their machines. One piece of software is an e-mail program called Mailtool which has a graphical mail reading program which allows you to make file attachments to mail messages.

I created a Mailtool version that comes out once a week. The message bodies are the summaries of the articles, and have attachments to the twenty or so articles for that week. That way people can quickly scan through the table of contents, find a story they're interested in, and go straight to that attachment to find the story.

K: Within each issue, have you worked out a particular strategy for organizing the information?

J: I discovered that because the information I post is very specific to Sun computers, sometimes I'll have five or six very specialized articles that I don't think are of general interest to people. I group those together so even those people who get the daily newsletter will sometimes get a collection of stories which even the daily people have to use the automatic response program. That way I can put out fairly esoteric details and not actually overfill people's mailboxes. If it's something that I think is of general interest, such as Sun announcing their quarterly earnings or a major new workstation I push that out to everybody.

K: How do you identify who gets what esoteric information?

J: I use a mailing list manager called Majordomo and operate separate mailing lists. The nice thing about Majordomo is that people can move themselves from one list to another without intervention. They can send e-mail to Majordomo, unsubscribe from one list and subscribe to the weekly list or the monthly list.

K: And you don't have to be involved?

J: The only time I get involved with Majordomo is if the address of the person making the change is different from the address to which the change is being requested. I'll scan all of the security features so people don't accidentally remove the redistribution aliases or improperly add their friends or remove their enemies or anyone else.

K: What about platforms other than Sun?

J: The Mailtool would be an example. However, some people liked the idea but didn't like that particular tool. In the time since Sun created the Mailtool program, a new Internet standard has evolved called MIME. MIME allows for the creation of attachments that can contain any subject matter, such as graphics or audio files, PostScript files or plain text, in any format PCs, Macs or other vendors' workstations.

K: With readers outside of Sun, do you have to adjust any sensitive content within the publication?

J: I have to be careful to implement some Sun policies or restrictions. One of which is how and where I distribute the U.S. prices. Basically they don't want me to share U.S. prices with non-U.S. customers. My program will check if there is a U.S. version of the article, basically with U.S. prices, and if the person making the request for the article is on the USA list, I'll send them the U.S. version. Otherwise, I'll send them the generic international version.

K: How many people are employed in producing this periodical and distributing it to the 150,000?

J: Three people. Me, Myself and I! I do this in addition to my full-time job as a systems engineer.

K: So you write it and edit it?

J: Well, I don't write it. The articles come from a variety of sources within Sun, particularly press releases. We have a series of internal product announcements called Sun Intros which are designed for the field people such as system engineers, sales reps, marketing people, not customers. I edit that and make it proper to give it to customers, which means sometimes changing the sales pitch, toning it down and removing internal contact information, and removing some sections which I don't think are necessary for customers. Some people within Sun ask me what makes me think I'm qualified to make those edits to internal documents. My job as a systems engineer means I represent the company to the customer so that's exactly what I'm qualified to do.

K: You do that every day?

J: Yes, and I think it's appropriate that after I make those edits, I send them back to product management people who originally authored them and explicitly seek their permission. I tell them this article will be read by hundreds of thousands of Sun customers and ask if they are happy with it. It also gives them the opportunity to make last-minute changes, like "We said it was going to go out in November, it's actually going out in December."

K: Or, "We forgot to mention this important feature, let us tell you about it." So, you're kind of like a meta-systems engineer?

J: Right, in fact, a large reason I started producing was it helped me locally, with my sales reps in my office and my local customers. There's a motto I like: "Think Globally, Act Locally." So, on behalf of my corporation I'm thinking locally, it's good for my local customers, but by acting globally, producing it in an electronic format which is inexpensively distributed throughout the globe to all my other fellow systems engineers, sales reps, customers, it helps them too with a little incremental effort from me.

K: You were doing the work anyway for your local office and now maybe you're even more motivated to do more?

J: Yes, for example, there's one group that produces a very nice printed newsletter once per quarter called Sun's Harvest. They have professional editors and colors and pictures and all the good stuff. I had them do the extra effort after they printed to extract the text and make an e-mail version of their newsletter. I probably distribute their information to more people than they do, at a fraction of the price.

K: That brings up an interesting point. We are often told there are twenty to fifty million people on the Internet, but those numbers actually apply to the numbers of people who have e-mail access.

J: That's right. I've had people refer to it as the core Internet. I've estimated that perhaps four or five million people can actually use Mosaic and FTP and compared them to the larger collection of networks which can exchange e-mails. There are perhaps 30 or 40 million of those people.

K: That would include the 1.5 million America Online members, the two million Prodigy members, the million CompuServe members?

J: Exactly. I've run across a number of bulletin boards overseas that want to make the connection and exchange e-mail. They have that capability, even with low-speed links. That number will always be greater than the number who have first-class access. Sometimes in the U.S. we forget just how inexpensive telecommunications are for us.

Elsewhere in the world it's even more expensive to have a telephone line, let alone the luxury of having local dialing for free. Where I come from, Scotland, if you're local dialing to the guy next door they charge you every minute, just like dialing long distance.

K: In Japan you need a $1,000 deposit to get a phone, and you have to wait a year.

J: So there'l always be more people with e-mail access. But in recognizing the fact that the Web has become so popular, over the last few months I've changed my production technique to one which I actually produce the articles in HTML, and make sure that the files look good in the Web format.

Then I process them into plain ASCII, which I format and distribute by e-mail. I place the actual HTML and plain ASCII files on an FTP server, which introduces yet another distribution methodology. Again, this is very useful for people in remote locations where they have slow-speed Internet links. They can connect in through FIP and download a megabyte file and get a whole week or month worth of articles, and then deal with them locally at their leisure.

K: Any plans to have a Website?

J: Well, I have my sneaky plan. Rather than having one Website, what I've done is placed all the HTML files on my FTP server and encouraged people to install them in their local web. I have four or five people who have done that already to my knowledge, and I think I'll see a lot more people do that as they realize these files are available.

K: You're really providing the raw material for excellent Websites for local resellers.

J: Exactly. I want to be the content provider rather than the service provider

K: Another important issue is cost. What kind of cost are we looking at to reach these 150,000 subscribers?

J: There are no real costs. Sun is already on the Internet, we have our links in place, we have our offices already set up. The sales office in Denver wants to redistribute my newsletter to their customers. I'll send one copy to them, then they can take care of paying the local phone bills, or the long-distance bills to customers there. As Sun's network has developed and the Internet develops, I can continue that model but to a large extent it's very inexpensive for Sun to provide that capability.

K: What about readers outside of the Sun loop?

J: Some of our customers pay to go on the Internet, so this is an added benefit to them being on the Internet. Some people use commercial services so they are paying a long-distance dial-up fee or a connect fee. So, there are costs to people out there, so for that reason, I also try to limit material to no more than 100 or so articles a month. That was also one of the reasons I went to offering the weekly version. That way people who are paying for it get just one article a week, and they only send for the articles which they think will be valuable to them.

K: So the menu also helps you cut costs for your readers?

J: Another interesting effect I've discovered is that if I look at the statistics from my automatic response program for a month, even though I've got close to 100 articles, the automatic response program may have sent out 450 or 500 different articles. So people still fetch articles up to a year old from the archives.

K: You know, even if there weren't bandwidth constraints, and someday we'll be in a world where there aren't bandwidth constraints, there are still human processing constraints. This idea of sending a menu and letting people look at what they're going to select is still an efficient means of delivery.

J: Given that my audience is relatively specialized, I can constrain all the types of information to be of interest to Sun users. As Sun users become more diverse, my present model may become difficult.

Sun users have very broad interests. There are developers, administrators, programmers, and their interests are different. It helps if you are an administrator, you can skip articles that are for developers and vice versa. Eventually, I may have to go to a model where I produce three newsletters as the amount of information I need to share increases.

K: What kind of software are you using to prepare the HTML you mentioned?

J: I use the standard text editor that comes from Sun, and have Mosaic and a WWW server installed in my machine so I can install a file and browse it with my own Mosaic viewer, make sure it looks okay. I also use an HTML editor called HoTMetaL to make sure that the HTML looks good on a browser. Afterwards, I have another window open where I use Mosaic to take the completed article and save it as formatted ASCII, then use another text editor. I kind of start off with text on the right-hand side and end with text on the left hand side, when I make sure that the post-processed version looks okay. Then I save those to my database.

K: Basically, you're running this whole operation off of one SparcStation. Your physical needs to do this are a Sparc workstation, the software that you've described, a connection to the Internet, time to develop your network of subscribers, and what else?

J: The mailing list management software I use, Majordomo, is publicly available. It makes it very easy to create multiple lists or have them more specialized. The automatic response program is homegrown; I didn't really find a program to do what I wanted it to do. I basically had to figure out how to process the incoming addresses and then provide users with a convenient syntax. In my case, that syntax consists of one or more article numbers, either on a subject line or on the body, and a few key words like help and index. I would like to expand that and offer more capabilities like search for example.

K: Currently, if I'm a subscriber, I'll get a list of articles with a number next to each, perhaps once a week?

J: You'll also get a paragraph about each so you can determine if you're interested. Then you can send e-mail to "flashback.com," my home machine with a list of one or more article numbers. Flashback's automatic response system will queue up your request, and process it immediately. I don't know how many requests I'm going to get at one time. By queuing them and having one program process them it has a predictable load on the machine, rather than posting a weekly summary and then getting 340 requests all at the same time, or having 340 response programs running all at the same time.

That's one of the challenges, I haven't seen many references to commercial automatic response programs because each person's requirements will be different. I had to figure out how to store each of the articles to make it easier for the program to figure out who sent for article 403, to see if that is a valid article, and then to reconstruct it and send it back to them. I guess that's one of the challenges for non-technical people doing this there isn't a lot of off-the-shelf software to help you do that.

K: You recently did a demographic study. How did that work?

J: I'm calling it a census. I've asked people to respond to the article I sent out, with the keyword PING on the subject line kind of similar to how people in the UNIX world check if another machine is up with TCP / IP. They can send a PING to check and see if that machine is up so I wanted to see which users are out there and simply respond so I can collect all the responses. The first day I had about 2,000 responses, in 48 hours I had 3000 responses. By the end of that week I noticed there were about 4,000 responses.

K: Can you tell which individuals these responses were?

J: I only get each individual's e-mail address which tells me a little about them. I strip off the peoples' names and have a list of the domain names. I can have some kind of where they're coming from, by country, and by their type of domain, such as "com" or "edu."

K: So, if I request a certain article are you able to record my request?

J: I record the time, the date, the article you requested and your e-mail address to produce monthly statistics as to which were the most popular articles. As publisher, I get feedback as to which articles people found interesting enough to send for. This helps direct me as to what kind of stories to send out. Most of Sun's press releases are interesting but sometimes we produce some that are so outrageously self-promoting that people aren't interested enough.

K: We were talking earlier about how you are filling a unique role. You're not marketing, you're not public relations, you're not a journalist. You're in a new area, but I guess it grows out of your job as a systems engineer.

J: The key thing is that I have to add value to the information I'm delivering. If it isn't the stuff that the customers are interested in, then they'll stop using the service

K: I understand you have another project underway.

J: Over the last couple of years I started carrying a lot of articles that weren't from Sun, but were of interest to the Sun community. I was a little unclear if it was appropriate for Sun to carry so much information about other companies without explicit permission. It just didn't feel right to have Sun posting all of this information for other companies. So, I started a separate endeavor called FlashBack which provides a mechanism for those same people to have access to a similar mailing list. I use the machine in my garage for both ventures. I can produce both newsletters in less time than it used to take to produce just one.

K: The audience is essentially the same?

J: They're all Sun people; that's the key thing I'm looking for. In the relationship with non-Sun vendors I act more as a user's advocate and make sure I can hold back the flood of advertisements. If you want to advertise, buy advertising space in one of the trade magazines. If you want more interesting stuff and more depth, you don't want an advertisement. The less the vendor slants toward himself, the more credibility he'll gain. I think people have learned that lesson.

K: But advertisements don't really take up space in an electronic publication. In fact, you probably really don't have to worry about size of material or space in the publication, do you?

J: It's electronic and it's relatively inexpensive in relationship to depth. If it's a printed magazine, the more pages, the more expensive it is. If it's a press release you get charged by the word, but if it's by e-mail, so many people use the fill-text-on-demand by e-mail capability that it's reasonable to have the articles fairly long. People will only send for them if they're interested in the subject. So, the type of information people typically put in a glossy brochure and hand out at a trade show would be a good candidate, such as the ones with more technical information, like a white paper.

K: What thoughts have you had about subscription fees?

J: I'm investigating a couple of different business models. One would be simply a fee per article or a sponsored article.

The problem with that approach is people need some evidence as to exactly how big my audience is and I haven't done a lot of that demographic kind of study. I'm looking at one in which I keep track of the leads that are generated and I charge on a per-lead basis. To make that work I'm thinking of an annual fee, to be listed in a vendor database, which is what people would see reference to in an article. I would post an article by a vendor, and for more information FTP article XYZ. Then people will send for article XYZ which would have the latest information by that vendor, including name, number, the fax number, their locations in Germany, Brazil, or wherever, and a list of all the other articles that vendor has posted.

By proactively contacting them once a quarter and validating that the information is correct, it wouldn't matter if someone fetched an article by that vendor and they had a change of location. When they send e-mail to get that vendor's contact information from the database, it would be right up to date. By keeping track of which people sent for the contact information, there would be a very high probability that those would be leads, and if I charge on a per lead basis with a ceiling, like $10 per lead up to $1,000 if there were 1,000 responses, they wouldn't be surprised with a large bill. It would only cost $1000. But also, the nice thing for them is if this is the wrong audience and there were no responses, it wouldn't cost them anything.

I think people are uncomfortable about actually trying it out. Plus, I would still retain the editorial rights to act as the users' advocate, and be able to say "You know, that's really not appropriate for this audience. It's really nice you sold lots of whatever to those people but that won't work here. Now, if you want to write a story as to why all those people bought all those things and the technology, then that would be interesting."

K: I realize you process a lot of material, but what kind of fact-checking or validation of vendor claims do you have to go through?

J: This is a very public audience, in a highly specialized marketplace, that spends a lot of money. People spend $5 billion a year on Sun computers. People spend probably three or four times as much on related software and hardware. That's a $20 or $30 billion per year marketplace. So, for a vendor to make a claim in public that they can't validate, this would be a bad idea. And the competition probably subscribes. That's one of the advantages of the Internet as Intel discovered recently, bad news travels very, very quickly.

K: You are a major information source in a $25 billion dollar market. In fact, you may not want to blow your own horn, but you've been meeting some interesting people by virtue of your publishing activities. For instance, who are you meeting with tomorrow?

J: Scott McNealy, Sun's chief executive officer. In fact, last month I published an article by Scott. I've been encouraging the Sun executives to recognize that they have a unique opportunity here. Sun senior officers can post articles that go directly to the customers on at least a monthly basis, without their own PR people. Without the newspapers and business being in the way, they'll find information slides straight to the users. I got very positive feedback from Scott's article.

K: From the readers?

J: Yes, the readers. I asked them expressly, "Here's an article from Sun's chief executive office, do you like having this direct stuff?" One of the things I'm planning on talking to him about tomorrow is the bi-directional quality of e-mail. That's to say, I can ask people for input or to suggest a topic, and ask for their feedback. Readers would get to tell the execs exactly what they think of Sun quality, software, hardware, or service. Then the executives would turn around and address those issues. Quality is a real good example for that because we've been investing a lot of money over the last year on quality and it's not really visible.

Actually it's a problem within Sun, when it takes a year or two for people to see the results. An executive can take some of the questions, gauge the audience interest and then turn around and respond within about ten days. When you think about it, there are few other media where people can do that. I mean it's not clear there are any! Trying to do that by direct mail, you're looking at a minimum of a twenty-day turnaround and big bucks. This is free.

K: I understand you were recently elected to the Board of Directors for the Sun user's group?

J: Part of the reason I was elected was a recognition of the contribution I've made to the user community. I've always tried to balance the fact that I'm a Sun employee and also a field employee and interact with customers directly. That's why those corporate marketing people sometimes are at a distance from the users and see things a little bit differently. At times I see myself, in some ways, as a user advocate too.

K: I think the whole project is a pro-users project. That also benefits the company.

J: To be able to apply my experience to highlight the important and shelter people from the normal day-to-day turnout and the mundane is an important part of my editorial contribution.

K: Have you seen other examples of people doing what you're doing and the way you're doing it?

J: There's a couple of commercial services that focused on high-performance computing. A company out of San Diego called HPC Wire has a fee-based service and produces only a weekly publication. The only way you can get an article is to send the article number to their automatic response program, like mine. But, you have to be a subscriber. They also have sponsorship mechanisms and advertising. They cover high-performance computing, such as supercomputers, massive parallels, and workstations too. I've made available their weekly table of contents to my subscribers because it's a large overlap in interest.

K: Do you think there is a lower-end market for such a publication?

J: I think the areas where there are high e-mail activity and high value make these services valuable. People are buying $55 games for their PCs, but it's not clear that a service like this would make sense in that area. Think about trade magazines for inexpensive PC products that tend to be measured in inches.

High-end magazines are narrower and highly focused. They tend to focus on expensive products with the hardware and the software that runs them. On the high end it's not unusual for people to spend $100,000 or $200,000 for one software package for one user, which is inconceivable in the PC world.

K: Do you see publication opportunities like this for other companies and other industries?

J: Any place where there's large amounts of changing information. You can select and share that information and apply proper experience to that filtering process. I think that's a problem that's going to grow and is growing exponentially. If people are attracted to the idea of using things like the World Wide Web (WWW), this is a real good way to get started.

People have been using desktop publishing for a few years now and discovered the biggest hassle in desktop publishing is the publishing part. The formatting, the layout, getting it to the printer, folding it, stapling it, that's all the work. The part that's the most fun, the most useful, and the most value-added is actually selecting content and editing the newsletter. All the trends that made the 80's the decade of desktop publishing will make the next decade the decade of online publishing.

K: What do you see as the earliest markets for this kind of publication?

J: The kind of people who make sense to market this kind of publishing system to today are in the high-tech industries. Other industries will start following. The government marketplace, for example is becoming highly connected to the Internet. Partially for their own purposes. This is a great way to satisfy the public mandate to make information available. But it works both ways. That means that the government is now accessible by e-mail, opening up many unexplored areas.

K: Perhaps the key point here is that basically one person, with one machine and one collection of software is operating a publishing venture that is reaching at least 120,000 people responsible for about $25 Billion of activity.

J: I think I have confidence in saying that I'm reaching a solid 10% of Sun's marketplace, a phenomenal number when you think about it.

K: I know you are busy with your own work, but do you ever consult with people who want to establish similar systems for their companies or their industries?

J: Not yet, but I would be interested in doing that. I think helping other people produce a similar publication would also help me continue my ideas.

K: It would be great if there were a community of publishers like you, you could all compare notes and share ideas.

J: I think one of the reasons I'm highly motivated to turn this into a commercial venture and to start making some money with it, is that I have more ideas than I have time to implement. I need people to help me. I would like, for example, to produce a real multimedia version. A reader could send for an article about a company and it would include PostScript documents, photographs or screenshots of a product, or maybe audio clips. This is all technology which is available today. It's easy to do.

Right now the Internet is like a printing press with paper and ink next to it. Everybody's standing around saying what a great invention it is. OK, fair enough, but it's time we actually started printing. That's the way I think it is for most multimedia. It's ready to start printing; we just have to start cranking it out. In the same way that printing is generic technology, this is the next generic communication technology

K: Thank John McLaughlin for sharing your experiences with us. How would someone get in touch with you to find out more about your publishing venture?

J: E-mail, of course. To flash@flashback.com or flash@sun.com.

The complete archive of Net Ventures can be read at:
www.HowtheWebWon.com

Appendix IV:

My Response to "What You Think You Know About the Web is Wrong"

Excerpt from: "What You Think You Know About the Web is Wrong"
by Tony Haile
Originally published by *Time Magazine*,
March 9, 2014

> *...As the CEO of Chartbeat, my job is to work with the people who create content online (like Time.com) and provide them with real time data to better understand their readers. I've come to think that many people have got how things work online mixed up.*
>
> *Here's where it started to go wrong: In 1994, a former direct mail marketer called Ken McCarthy came up with the clickthrough as the measure of ad performance on the web. From that moment on, the click became the defining action of advertising on the web. The click built huge companies... and promised a whole New World for advertising where ads could be directly tied to consumer action... [end excerpt.]*

***

I've never had the pleasure of meeting Tony Haile, but I can say that Chartbeat was not, and is not, a lightweight company. They provide real-time web analytics to "over

313

1,000 media brands across 70 countries" and their clients include the *New York Times*, the *BBC, Hearst, Gannett, Warner Brothers*, and *Paramount*. So presumably Haile is a reliable source for "the fact" that I came up with "the click-through as the measure of ad performance on the web". Given that attributing this to me was not meant to be complimentary ("Here's where it started to go wrong..."), he was hardly intending to do me any favors by giving me the credit.

I had four reactions to seeing this article which I could have easily missed if someone hadn't pointed it out to me:

1. My first reaction was: "Little old me? I'm the guy who came up with the metric that companies like Google and Facebook have built their fortunes on? Gee, and they've never even sent me a Christmas card."

2. My second reaction, as soon as I reminded myself that gratitude is better than sour grapes: "Thanks for noticing."

3. Then, because sour grapes are not the worst thing in the world, I thought of sending him this note and (I still might):

"Tony,

It was 2014 when you wrote this. It's 2024 now.

If clicks, and therefore the selling of clicks, are not the defining action of advertising on the web, don't you think by now that the collective wisdom of Silicon Valley, Madison Avenue, and Wall Street, with trillions of dollars at stake, would have come up with something better? I didn't impose my observation on the world. I merely was the first to state something that's now obvious. By this logic, the first person to notice that water was wet (there had to be someone) would also be the cause of everyone who slips on a freshly mopped floor or drowns in the ocean.

In any event, long before 2014, the smart people I know in the selling business (which is different than the advertising business or the business of selling web metrics to publishers) were tracking clicks all the way through to sales and we've been doing it for years. If your distinguished clients (the *New York Times*, the *BBC* et. al.) have allowed chasing clicks to pervert their editorial policies, that's on them."

4. But the most outrageous thing of all is on Tony's personal website. He reports that in 2006 *New Woman Magazine* voted him one of the 10 sexiest men on MySpace. I was never even given the chance to nominate myself. While I'm sure that there's a statute of limitations, there must be a higher authority I can appeal to.

Endnotes

***

A (p.10)

'Electronic Brain' Installed Here

Francis W. McCarthy (far left). My Dad got involved with computing in the 1950s. This picture of him was from before I was born. He worked a lot. I didn't see him much.

***

B (p.153)   This last item was a long-time interest of mine. In preparing for this book, I found some correspondence between me and Arthur C. Clarke from the mid-1980s. I was vacationing in a little Mexican fishing village (Barra de Navidad, Jalisco) waiting to make a long-distance call at the local "casesta de large distancia". In those days, especially in small towns, if you wanted to make a long-distance call, you'd have to hoof it to a special facility to do so and wait on line (not online, literally "on line") until it was your turn to make a call.

The experience reminded me of Clarke's 1945 prediction that satellites in geostationary orbit could theoretically provide worldwide radio coverage (remember this was before TV). I also knew Clarke was interested in human progress so I faxed him from that little village. I wondered if there might not be some way to use a system of satellites and solar-powered PCs to beam down farming, engineering, and medical information to villages around the world that had little access to modern technology. He liked the idea.

In 1992, soon after I was first exposed to the idea of multimedia, I sketched out a combination phone, computer, and fax machine, which would use existing technology - the phone system for access to a repository of documents, a personal computer to scroll through them, and a fax machine to print out the pages you wanted. This was a year before I realized there was such a thing as the online world.

I improved on the idea and made it a touch screen. (See photos on next pages.)

This is from a notebook that I started after being exposed to the idea of multimedia, but before I went to ONE BBSCON '93. I envisioned a touch screen on a desktop with a display of classified-ad-looking entries. Pushing a button would call up a specific video. If you needed to print something off the screen, you could order it to be sent to you via your fax machine.

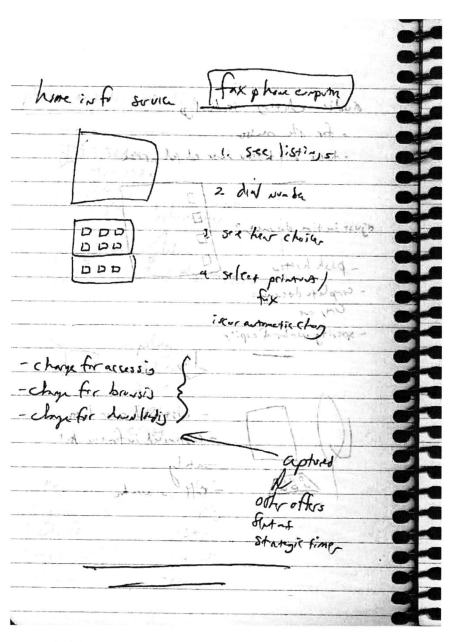

I called this one "The Home Info Service." The idea was to link your phone with your computer and your fax machine. Revenue would come from a monthly access charge, a charge for "browsing" (I think I meant access time), and a charge for each document you downloaded.

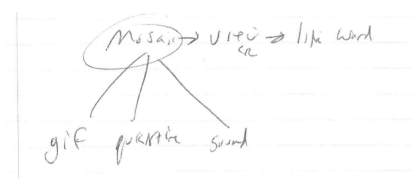

After ONE BBSCON '93, I attended as many Internet meetings as I could find. On December 20, 1993, I finally got my first look at the Web. It was a small part of a long presentation about all the technical bells and whistles available on the Internet. This is what I jotted down about it. Yes, it took four months, even in San Francisco, for a highly-motivated and reasonably intelligent person to find a way to get a glimpse at the Web.

***

C (p.155)   To give you an idea of where things were at at the time of this announcement, this specialized business newsletter, which had a list price of $279 a year, felt it necessary to include the following instructions to accompany an article about the Internet Shopping Network (ISN), an early online store, in the same May 1994 issue:

> For information on ISN, send E-mail to info@ internet.net. Those with Mosaic can enter the ISN storefront by using Mosaic's URL. (Universal Resource Locator) and typing http://shop.internet.net.

In this same issue, they introduced their readers to a brand new term: "Web Master".

In another article in this issue, they referred to firms that offer to "mount marketing information on the Internet" for clients and quote an expert saying, "There's probably (only) a dozen companies that have any hope (of surviving by offering this service to clients)."

▼右がヘンリー・ダーキン。知る人ぞ知る大金持ちである　　　　G.C.

暮らしを送っている。自

の道楽ともいえるの

リーのような"エンジェル"（起業を

支援する個人投資家）がたくさんい

間に合わな

アップル

代初　この　いた　ンタ　再び　つつ　プル　チャ　れな

A few years later, when I briefly became "famous in Japan" there was a steady stream of journalists from that country knocking on my door to do stories. I always brought them to 3220 for the photo shoot and I'm so glad I was able to get Henry to pose with me for this photo. I can't possibly overstate what a wonderful human being he was.

Lest you think that the Internet Business Report was published by a fringe group out of touch and behind the curve of the computer industry, it was a publication of CMP Media, which published computer industry bibles of the 1990s like Computer Reseller News, Computer Retail Week, and VARBusiness. This is a good barometer of the mainstream computer industry's awareness and appreciation of the Web in May of 1994.

One last bit from this issue. According to the article on Clark and Andreessen forming a new company, the original name of the company was Electric Media Company before it came out at Mosaic Communications. It subsequently changed its name to Netscape.

***

D (p.166)   Henry came from a pioneering California family that arrived in the state in the 1840s. They were accomplished entrepreneurs and generous philanthropists. His maternal grandmother developed a spectacularly successful citrus farm. Another project she developed ended up becoming the grounds of Claremont Graduate University. His father and brother founded Dakin, Inc., the toy Company which for a time was the nation's biggest maker and marketer of quality plush toys including teddy bears. One of their biggest hits was a Garfield the Cat with suction cups for paws. For a while in the early 1980s, it seemed to be on the inside of one out of every ten cars as well as in many homes. In its peak year, they sold $50 million worth.

Henry was a scientist, not a businessman. He was trained in science at Harvard and received a degree in electrical engineering from them in 1958. In the 1960s, he worked at Lawrence Berkeley Laboratory where he developed a pocket radiation detector that is still commonly in use today. He was one of the very early users of computers to publish books, newsletters, and posters, now a common everyday practice that later came to be known as desktop publishing. One of his projects in this area involved supporting underground newspapers in the Soviet Union ("Samizdat") and translating and publishing writing smuggled out of Soviet political prisons. At one point this activity earned him the designation "the most dangerous man in America" in an article in Pravda. He also hosted video teleconferences - in the 1980s(!) - between Russians and Americans using a satellite dish on 3220's roof.

***

E (p.180)   In May of 1994, GNN won the "Best Commercial Site" and numerous other awards at the First International Conference on the World Wide Web held in Geneva, the home of CERN and where the Web had been born.

Robert Cailliau, the co-inventor of the Web, organized it. He recalled that approximately 700 to 800 developers showed up, but only 400 could fit in the space he rented. When they put on a second conference later that year in Chicago in October, 1,300 people attended. Note that the attendees were software engineers, not business people, and not entrepreneurs.

Though the Geneva event has "First" in its name, the very first gathering of web developers was held in July of 1993 in Cambridge, Massachusetts where Tim Berners-Lee was teaching (MIT). A grand total of 30 people showed up. One of them was Marc Andreessen and it was the first time Andreessen and Berners-Lee met.

The organizer of the conference was none other than Dale Dougherty who was in the final leg of getting Global Net Navigator ready for its launch a month later. One of the eventful things that happened at this meeting was Marc Andreessen's advocacy for a new HTML tag which he'd originally proposed on February 25, 1993, saying:

I'd like to propose a new, optional HTML tag:

IMG

Required argument is SRC="url

***

F (p.182)   *"Marc,*

I was explaining Mosaic to a guy who is a VP of a very well-known advertising agency. He was fascinated.

I was surprised he hadn't heard of it, but then I remembered how conservative the ad agency business is...

Then it occurred to me that ultimately all of the biggest users of Mosaic will be advertising agencies just as they are among the biggest users of print, film, and other media.

Wouldn't it be interesting if the people who are designing the tool and the people who are going to be using it could get together now and learn from each other? Could speed the evolution of things significantly."

His answer:

"Hi Ken,

Absolutely agree with everything you say. Some agencies are more advanced than others, though. Did you read about the chairperson of Procter & Gamble's speech to the Amer Assoc of Ad Agencies a couple of months back?"

***

G (p.184)   In the 60's, Jim Warren was General Secretary of the utopian Midpennisula Free University, one of the largest and most successful of the many "free universities" that sprung up in that era. For $10 anyone could get the course they offered list-

I got this picture from Dan Kottke. He's the guy on the right. That's Steve Wozniak on the left and Steve Jobs in the middle. The year was 1977.

ed in their quarterly newsprint catalog and teach on any subject they wanted Campus-less, Free U classes were held in homes, storefronts, and parks.

From that, he dove head first into computers and was a busy freelance programmer and consultant in the era right before personal computers and also became a leader in local computer industry professional groups. In 1977, noting the burgeoning interest in personal computers, he started the West Coast Computer Faire, the spelling of "faire" having been inspired by the Renaissance Pleasure Faires which like free universities were also a creation of the 1960s.

The first Faire was held at San Francisco's Civic Auditorium on April 16-17, 1977. Sponsors included the Homebrew Computer Club, then the second-largest computer club in the U.S. Apple bought a booth which wasn't much more than a card table. The ad for the first Faire stated that "San Francisco Bay Area - Where it All Started - Has its First Home Computing Convention" and promised "7,000 to 10,000 people, 100 conferences, 200 Commercial and Homebrew Exhibits".

Warren only expected 7,000 attendees at most, but 12,000 showed up. People waited in line for hours to get in. "We didn't know what we were doing. The exhibitors didn't know what they were doing, and the attendees didn't know what was going on but everyone was excited and congenial and undemanding," he told Steven Levy, author of *Hackers: Heroes of the Computer Revolution* (1984). In 1983, he sold the Faire for $3 million to Prenctice-Hall who sold it to Sheldon Adelson, the owner of COMDEX which supplanted the Faire as the biggest personal computer conference in the world.

***

H (p.187)   There had been a handful of meetings, large and small, about the Internet in general and some included panels on business issues intermingled with a tech-heavy agenda. Alan Meckler of Mecklermedia put on the biggest one. It was called Internet World '93 and was held on October 25-27, 1993, in the Moscone Center in San Francisco, one of the city's largest convention venues.

I attended this event with high hopes. Two months earlier, I'd been introduced to the Internet at ONE BBSCON'93 in Colorado and I was hungry for more information, specifically about how people were using the Internet for business. The conference didn't add much to what I'd already picked up in Colorado which is to say not much was happening on the business front.

Here is the list of featured seminar topics from the East Coast Internet World which they ran two months later. It shows programming that was appropriate for the time and in some cases even visionary, but topics of interest to entrepreneurs were few and far between.

Document Delivery and the Internet, Basics
for Small Institutions, Navigating the
Net, Leasing & Copyright, Government
Information on the Net, Getting Connected,
Market-based global Libraries, Wireless
Access, Electronic Texts, Doing Business on
the Internet, Cost Justifying the Internet,
NREN and Publishers, Fax Publishing,
Publishing on the Internet, National
Networks, Protocols and Standards, Retrieval
Tools, Community Networking.

The full name of the conference was *Internet World '93* and *Document Delivery '93* and was subtitled "Conference and Exhibition Dealing with the Commercial and Non-commercial Utilization of these Services and the Impact they have on Information Providers and Users"

Meckler spent the first twenty years of his entrepreneurial career producing seminars and publishing books and trade journals for university and other specialized libraries. In the very early 1990s, thanks to his contacts with librarians, he got his first look at the Internet, pre-World Wide Web.

Two years before Internet World '93, he'd been following his formula of finding information niches neglected by other publishers and building newsletters, directories, and seminars around them. Initially, he viewed the Internet as a special topic of interest to librarians only and started a niche newsletter *Research and Education Networking* with librarians as his intended market. In the process of developing that newsletter, his perspective changed dramatically, and in 1993, he came to the conclusion that the Internet was going to be an even bigger advance than the personal computer.

By the end of 1994, his magazine *Internet World* had 70,000 paid subscribers, peaking at 125,000. The Internet World Conference spread to 33 countries in addition to the U.S. In 1994, he started MecklerWeb.com which later became Internet.com, an integrated network of over 50 websites, most dedicated to electronic and online industry issues. In 1998, Penton Media bought the company he founded for $274 million, with Meckler retaining 80% of Mecklermedia.com website.

***

I (p.198)   I called it *the* TED conference because until 2008 there was only one TED conference (Technology / Entertainment / Design) held each year starting with its founding in 1994 by Richard Saul Wurman and Harry Marks (1931-2019), and it had always been held at the Monterey Conference Center in Monterey, California. Marks was the godfather of the broadcast graphics we now take for granted. Wurman, trained in architecture, was a master practitioner of information design and has published over 100 books on a wide range of topics including many of the iconic ACCESS series which included travel guides.

***

J (p.200)   After AT&T was broken up in 1984, AT&T kept Bell Labs, renamed AT&T Laboratories. AT&T's monopoly on telephone service in the U.S. generated enough revenue to maintain an extraordinary quantity and quality of thinking, basic research, and experiments with technology. Think Xerox PARC on steroids. Everything that appeared in the ads in 1993 was being worked on at Bell Labs when the ad series ran.

In the mid-1980s, what turned out to be a prototype of this campaign was developed and screened at Disney's EPCOT Center. It proved to be a hit, especially among young people. At the time, Japanese consumer electronics companies like Sony and Panasonic were all the rage and AT&T was coming to be perceived as a has-been. The "You Will" TV campaign was seen as a remedy to that.

The director of the spots was David Fincher who was 31 years old at the time. He co-founded Propaganda Films in 1986 at the age of 24. Among other things he directed Madonna's "Express Yourself" and "Vogue" and went on to direct feature films like *Alien 3, Fight Club, Panic Room, The Girl with the Dragon Tattoo, and The Social Network*. Early in his career, he became a visual effects producer and worked with George Lucas and Industrial Light and Magic. Coincidentally Fincher's family had been a neighbor of George Lucas when the family moved from Denver to San Anselmo when Fincher was two years old.

The creative team on the ad, produced by the ad agency N.W. Ayer, was Copy Supervisor Gordon Hasse, Art Supervisor Nick Scordato, and Producer Gaston Braun. Tom Selleck narrated. Robin Williams auditioned, but apparently, his voice wasn't a good fit.

AT&T spent $50 million on the campaign and in addition to TV it also included radio and print - and later the Internet. It was one of the biggest branding campaigns of its time. I've devoted so much time relaying this history because it no doubt was seen at least once, or more likely many times, by the people working on Internet and multimedia projects in 1993 and must have been a powerful inspiration, even if just a subliminal one.

***

K (p.201)   Modem Media's first project was trying to sell online shopping malls to pre-Internet online computer services companies like Delphi, Dow Jones, and GEnie. They connected with GEnie and designed and managed the GEnie Mall for them. This venture may well have been the first use of e-mail as a marketing tool to sell products. An unethical employee went behind their backs and offered GE a better deal which GE, unethically, took.

Despite this early setback, the company survived and thrived and when the Internet rolled around and big agencies were looking to subcontract specialized digital work, Modem was on top of the go-to list thus N.W. Ayer hired them to do the AT&T banner ad. Joe McCambley and Craig Kanarick did the actual design.

Modem Media became one of the key players in the early years of advertising on the Web. They were so important that when DoubleClick, the company that introduced online ad serving and targeting was getting started, one of their biggest initial challenges was getting Modem Media's attention. To do this Kevin O'Connor, who co-founded the firm with Dwight Merriman, targeted the Modem Media domain with ad messages congratulating CEO O'Connell on the recent birth of his twins. Targeting ads with that level of precision was unprecedented at the time, baffled Modem Media's staff, and got Modem Media's attention. In 2008, Google bought DoubleClick for $3.1 billion dollars.

***

L (p.209)

They finally started to get it. April 1995: "Braving one of the last ferocious winter rainstorms of the season, a crowd of between 200-250 International Interactive Communications Society (IICS) members sloshed their way to Cupertino... to hear Ken McCarthy, publisher of the Internet Gazette."

***

M (p.217)   This might be a good time for me to comment on the current AI craze. In 2023, I published a short book on the subject called *The Artificial Intelligence Question and Answer Book*. Writing a book does not make me an expert, but I have given more than a minute of thought to the topic.

First, in the 1990s, all that companies had to do was add a ".com" to their business name - whether they had anything to do with the Internet or not - and it would yield an instant uptick in their stock prices, free PR, and a flood of new investors. The same thing is happening again with AI.

Second, many things that are being called "AI" are nothing of the sort. To quote Arvind Narayanan, a professor of computer science at Princeton University on the subject of predictive AI: "The shocking thing to me is that there has been virtually no improvement in the last 100 years. We're using statistical formulas that were known when statisticians invented regression...but it (predictive AI) is sold as something else."

Third, the Turing Test, which was proposed by Alan Turing in 1949, the idea of which is as follows: A computer can be said to have passed the Turing Test (which he originally called the "imitation game") if, in a text conversation, a human evaluator could not reliably tell the computer's answers apart from ones he'd get from a human being. The original definition of success specifically excluded the need for the computer to give accurate answers.

The standard was how closely the answers from the computer *resembled* answers from a human being. I think anyone over the mental age of five recognizes the problem here. Is it really a crowning achievement to develop a technology that delivers blatantly incorrect information in response to questions, but does it in a way that seems to be coming from a reliable human-like source?

Fourth, and related to ChatGPT's disclaimer: "Chat GPT can make mistakes. Check important info". What exactly is the difference between "important" and "unimportant" information and who decides? If you are posing as an information source, isn't the whole reason for your existence to dispense *accurate* information? If I have to check on everything you say because a significant percentage of everything you tell me, even about simple matters, is completely wrong shouldn't I be looking elsewhere for information?

Fifth, and related to the above, completely inaccurate answers like the ones I received on the early history of the Web, given with supreme confidence, can and I'm sure will find their way into school papers, articles, broadcasts, and even books. This will cause a degradation of the quality of information, not an improvement.

Sixth, from the numbers I've heard, as of now, AI has attracted $1 trillion in investment and has yielded $30 billion in revenue (revenue, not profit). I appreciate that world-altering things can and have come from initiatives that make little or no money on the front end, but $1 trillion is a big number even today and a 3% return (revenue, not profit), troubles the part of me that can do arithmetic.

Seventh, I'm frequently told by enthusiasts, "Yes, now that you mention it, ChatGPT has some problems, but AI is doing amazing things in other fields like medicine." I could write a book on all the "amazing" things modern medicine has done for us. (In fact, I wrote one, *Fauci's First Fraud*.) I am less than amazed at the current state of medicine, the problems of which have nothing to do with a lack of technology. I don't think AI has a cure for the problems of our medical system which boil down in no particular order to arrogance, corruption, greed, and willful and abysmal ignorance about basic science.

Eighth, the electricity needed to carry out the computational de-
mands of AI (which apparently includes serving up inaccurate
info on a silver platter) is not a trivial matter. In the 1990s, I was
confident that the telecom challenge of getting a high-speed connection into everyone's homes, schools, and offices was attain-
able. I am not at all confident that the challenge of developing
the new power sources and electrical grid improvements needed
to power AI s projected growth is possible. Also, given that the
only idea anyone has put forward on how to do this, other than
building a small nuke plant for each new AI data center, which
is an actual proposal on the table, is that somehow AI will give
us "smart energy" and the real world, physical problems will
magically disappear. And pigs will fly. We shall see. (All that
said, I use ChatGPT daily and find it increasingly useful.)

***

N (p.230)   There's a fatal flaw with Anderson's concept of long
tail, "selling less of more". In 1994, I coined the term "the lifeboat concept" to describe the dynamics of creating and marketing content. You
can find it in the transcript of my November 5, 1994 talk at the end of
this book, Appendix II. I pointed out that the Internet vastly increases
the size of what I called the content "lifeboat," the range of content that
can be profitably sold under the old media system. Thanks to the Internet, massive amounts of content that would have never otherwise seen
the light of day suddenly became financially feasible to market. I think
by any measure this is a better model for companies to focus on versus
Anderson's "sell less of more" strategy of having a vast inventory of
very low-demand items that in the aggregate *might* make a profit.

***

O (p.252)   Just 25 years after the launch of the substandard
and unimpressive Windows '95, Bill Gates enjoyed a new incarnation as a medical and public health genius, a role he purchased
by directing the Bill and Melinda Gates Foundation to give away
$319.4 million in grants to news outlets all over the world to

write stories about science and public health. He was also the #2 financial supporter of the World Health Organization (WHO), second only to the United States. This was the institution whose director, a non-physician and non-scientist, declared COVID a "global pandemic." The very early and prominently publicized virus transmission and death rate projections that were so unhinged that they later had to be retracted, came from the Imperial College of London and the University of Washington, two institutions Gates was a major financial supporter of. Neil Ferguson of Imperial College of London has made numerous grossly exaggerated projections in the past.

The news media went along with Gates' fiction that the entire world population was at risk from a deadly, untreatable disease caused by a new virus that only he and his business partners had the remedy for, namely mRNA vaccines. Gates led the charge calling for mandatory digital vaccine passports and coercing large numbers of people (employees, students, hospital patients, and others) into receiving vaccines that he had a massive personal financial stake in.

When the VAERS (Vaccine Adverse Event Reporting System) data started coming in showing that there were more injury reports from these new vaccines than all other vaccines administered since the founding of VAERS in 1990 combined, Gates started backpedaling on his media pronouncements, disappeared from the public eye after having been on television constantly, and sold the shares he'd acquired in Moderna, the mRNA vaccine marketer, near the top of their sales price.

From the economic carnage suffered by tens of thousands of locally owned small businesses to the unprecedented rate of childhood suicides, to the abusive treatment of citizens by every level of government, to the breaking of critical supply chains, to the rate of reported vaccine injuries exponentially higher than ever seen in the history of medicine, to the moral failure of the medical profession to protect its patients from unproven, ineffective, and dangerous treatments, the COVID crisis was the greatest social catastrophe of the 21st century, and one would be hard pressed outside of war, to find one comparable in the last 100 years. Without Bill Gates and his multivarious "contributions" to the crisis, it's hard to envision any other path by which such prolonged official madness - supported by the news media and enforced by government at all levels - could have happened.

***

P (p.260)   Of all the financial assets (receivables other than taxes) that the U.S. government owns which one is the biggest? Student loans.

They amount to 38% of the federal government's total financial assets. That's eight times larger than the total outstanding mortgages it holds for a total of $1.73 trillion. This loan balance increased well over 100% since the bank-created financial meltdown of 2008. It appears that in order to create receivables for a debt-swamped government budget, the student loan spigot was turned on full blast. What have universities done with the money? Your guess is as good as mine.

Image Credits

All images from the collection of Ken McCarthy, except:

Cover of Ken McCarthy's *Internet Business Manual* Bunksha (Tokyo). (Page 207)
Used by permission.

Photo of Steve Jobs and Steve Wozniak, from the collection of Dan Kottke. (Page 326)
Used by permission.

Photo of Ken McCarthy and Henry Dakin, from *The Diamond Weekly*, Tokyo: October 6, 1996. (Page 322)
Used by permission.

Index

Numbers

A

B

Match.com, 229

McCambley, Joe, 331

McCarthy, Ken, 5, 189, 272, 294, 313, 317, 331, 337

McCool, Rob, 148, 156, 161

McLaughlin, John, 235, 291, 293-311

Meckler, Alan, 327-329

Mecklermedia, 327, 329

media buyers, 132-135, 164, 172, 177, 197

Media Lab, 111-112

Microsoft, 2, 17, 105, 109, 136, 155, 162, 212, 241, 252-256, 260-261, 268

Microsoft Network (MSN), 253-255

MicroTimes Magazine, 95, 137

Mittelhauser, Jon, 148, 161, 194

Mizel, Jonathan, 227

Mobile Marketing Handbook, 237

Modem Media, 201-202, 217, 330-331

modems, 54, 92, 101, 129, 175, 201, 280

Montgomery Ward, 122-123

Montulli, Lou, 161

Morse code, 141

Morse, Samuel, 1

Mosaic, 133, 143-144, 147-149, 156-158, 160, 162-163, 172, 176, 182-184, 187, 188-191, 194, 243-245, 275, 299, 302, 321, 323, 325

Mosaic browser, 149, 157-158, 172, 176, 190, 243-244

Mosaic Communications, 156, 182, 244, 323

MSN (Microsoft Network), 253-255

Mueller, Bettina, 47, 52, 59, 62, 65, 74, 78-80, 82, 93, 106, 176, 213-215, 221, 224

multimedia, 105-107, 109-110, 112-113, 116-119, 121, 131, 136-137, 152, 165, 182, 183, 187-188, 204, 209, 218, 229, 251, 275, 276-277, 311, 318-319, 330

multimedia CD-ROMs, 105, 112-113, 136, 152, 204, 251-252, 254, 275-277, 286

multimedia developers, 106, 118, 204, 275

MySpace, 315

N

Narayanan, Arvind, 332

NASDAQ, 204, 219-220

National Center for Supercomputing Applications. See NCSA

National Science Foundation Network (NSFN), 145

NCSA (National Center for Supercomputing Applications), 146-150, 160-161

Netscape, 121, 155, 162, 182, 184, 187, 190-191, 203, 210-212, 215-216, 218, 221-222, 229, 242, 244-247, 249, 253-256, 323

Netscape IPO, 212, 246, 249, 253, 256

Newsom, Gavin, 208

Niehaus, Ed, 215, 252

NSF (National Science Foundation), 93, 145, 149

O

O'Connell, GM, 201, 217, 331

O'Connor, Kevin, 331

Ogilvy, David, 65, 72, 215

O'Keefe, Steve, 64

ONE BBSCON'93, 86-89, 91, 95, 100-
102, 116, 119, 131, 136, 184,
319, 321, 327

OnRamp, 202

Oracle, 136, 224

O'Reilly, Tim, 180, 247

Oslin, George, 107-109, 140, 164

P

Pac Bell, 189-190, 275

Paine, Thomas, 115

Pandora, 165

Parker, Jeannine, 183, 187

Parker, Sean, 262

Patel, Avani, 204

pay-per-click, 230-231, 233, 237

PeaceNet, 95, 136

petabytes, 216

Polhemus, Guy, 27, 32

Princeton University, 11, 15-16, 77,
81, 214, 259, 332

Procter & Gamble, 163, 232, 325

Prodigy, 86, 124, 251, 255, 299

R

Radio Shack, 10

Rickard, Jack, 100, 280

Rosenthal, Sandy, 209

Rudl, Corey, 229

S

Sackheim, Maxwell, 65, 72

San Francisco, 5, 42, 43, 44, 52, 53, 55,
59, 60, 61, 65-66, 77-78, 81-82,
101, 104-106, 109, 112-113,
115, 121, 122, 124, 130, 132,
136, 137, 139, 155, 164-165,
181-183, 188, 204, 206, 208-
209, 214-215, 221, 223-224,
226-227, 229, 251-252, 256,
275, 321, 326-327

Schmidt, Eric, 260

Schwab, Klaus, 250

Schwab, Victor, 65

Schwartz, Eugene, 12, 65, 215

Scientific Advertising, 267

Scully, John, 252

Sears catalog, 288-289

Selver, Charlotte, 47, 52

Shannon, Claude, 106

Siino, Rosanne, 162, 164, 183, 210

Silicon Graphics, 150-151, 157, 162,
244

Silicon Gulch Gazette, 185

Silicon Valley, 104, 150-151, 156-157,
161-162, 182, 223, 229, 244,
247, 251, 253, 256-257, 314

Simon, Doug, 217

Simpson, Homer, 189, 243

Slip.Net, 138, 186

Smarr, Larry, 149

software industry, 122, 136

How the Web Won
Online Archive

www.HowTheWebWon.com

The following video, audio, and text can be found at the *How the Web Won* website:

* Video of Ken's Nov 5, 1994 talk, "Which Way the Web?"

* Marc Andreessen introduces Mosaic. Video. Nov 5, 1994

* Audio interviews with Internet pioneers Mark Graham, Steve O'Keefe, and Rick Boyce

* Values: Interviews with Jim Warren, Ronald Gross, and others

* Ken interviews Jonathan Mizel, online marketing pioneer (2024)

* Ken interviews Corey Rudl the original Internet marketing wunderkind (1999)

* Net Ventures newsletters Volumes 1-3 (1995-1996) E-mail publishing, website design, selling software online

* What modems sounded like in the 1990s

www.HowTheWebWon.com

Also by Ken McCarthy

The System Club Letters

Testing: The Key to Direct Marketing Profits

The Artificial Intelligence Question & Answer Book

Death, Resurrection, and the Spirit of New Orleans

Piacenza: A Hidden Jewel On The Crossroads Of History

Fauci's First Fraud: The Foundation of Medical
Totalitarianism in America

What the Nurses Saw

The Nuremberg Code: 75th Anniversary Commemorative
Edition

John F. Kennedy: Anti-Imperialist

Bank Failures, Bank Runs, and Financial System Collapses